D1128407

Handbook
of
Writing
for the
Mathematical
Sciences

Second Edition

Handbook
of
Writing
for the
Mathematical
Sciences

NICHOLAS J. HIGHAM

University of Manchester
Manchester, England

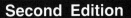 Society for Industrial and Applied Mathematics

hiladelphia

Library of Congress Cataloging-in-Publication Data

Higham, Nicholas J., 1961-
   Handbook of writing for the mathematical sciences / Nicholas J.
   Higham. -- 2nd ed.
   Includes bibliographical references and indexes.
   ISBN 978-0-898714-20-3
   1. Mathematics--Authorship. 2. Technical writing. I. Title.
   QA42.H54 1998
   808'.06651--dc21   98-7284

siam is a registered trademark.

Contents

Preface to the Second Edition

In the five years since the first edition of this book was published I have received numerous email messages and letters from readers commenting on the book and suggesting how it could be improved. I have also built up a large file of ideas based on my own experiences in reading, writing, and editing and in examining and supervising theses. With the aid of all this information I have completely revised the book. The most obvious changes in this second edition are the new chapters.

- *Writing and Defending a Thesis.* Since many of the readers of the book are graduate students, advice on how to write a thesis and how to handle the thesis defence was a natural addition.

- *Giving a Talk.* The revised chapter "Writing a Talk" from the first edition gives advice on preparing slides for a talk. The new chapter explains how to deliver a talk in front of an audience.

- *Preparing a Poster.* The poster is growing in popularity as a medium of communication at conferences and elsewhere, yet many of us have little experience of preparing posters.

- *$T_{\!E}\!X$ and $\mathrm{\LaTeX}$.* Since the first edition of this book was published, $\mathrm{\LaTeX\,2_\varepsilon}$ has become the official version of $\mathrm{\LaTeX}$, thereby solving many of the problems involving, for example, incompatible dialects of $\mathrm{\LaTeX}$, font handling, and inclusion of PostScript figures in a $\mathrm{\LaTeX}$ document. I have moved the discussion of $T_{\!E}\!X$, $\mathrm{\LaTeX}$, and their associated tools to a new chapter. Many more tips on the use of $T_{\!E}\!X$ and $\mathrm{\LaTeX}$ for typesetting mathematics are now given, and the discussions of $\mathrm{Bib}T_{\!E}\!X$ and indexing have been expanded. The many mathematical symbols in the AMS fonts have been added to Appendix B ("Summary of $T_{\!E}\!X$ and $\mathrm{\LaTeX}$ Symbols").

Among the new material in existing chapters, the section "How to Referee" in the chapter "Publishing a Paper" offers advice on this important aspect of the publication process, and in the chapter "Writing a Paper" suggested formats are given for referencing items on the World Wide Web.

The renamed chapter "Aids and Resources for Writing and Research" contains a new section "Library Classification Schemes", which should help readers to find their way around libraries. The material on the Internet in this chapter has been completely rewritten in the light of the World Wide Web (which was not mentioned in the first edition). I have minimized the number of URLs (Web addresses) given, for two reasons. First, URLs can and do change over time. Second, if you want to know more about almost any subject mentioned in the book, just choose appropriate key words (e.g., "mathematical writing", "Roget's Thesaurus", or "Society of Indexers") and invoke your favourite search engine from your Web browser. There is a good chance that you will find the information, or particular Web pages, that you are looking for.

The subject of mathematical writing can easily become dull and boring, so to liven it up I like to include anecdotes, unusual paper titles, humorous quotations, and so on. The new edition contains many more of these.

Separate author and subject indexes are now provided. The author index removes some clutter from the subject index, and you can use it to find where references in the bibliography are discussed.

The bibliography has been updated. Many new editions of books are referenced and over 70 new references have been added.

A Web page has been created for the book, at

`http://www.siam.org/books/ot63/`

It includes

- Updates relating to material in the book.

- Links to references in the bibliography that are available on the Web.

- Links to other Web pages related to mathematical writing, LaTeX, BibTeX, etc.

- Links to Web pages giving examples of posters.

- The bibliography for the book in BibTeX form, which is also available from Bibnet as `han-wri-mat-sci.bib`.

Several people helped with the second edition by reading and commenting on drafts:

David Abrahams, Henri Casanova, Bobby Cheng, Tony Cox, Des Higham, Doris Higham, Nil Mackey, Alicia Roca, Françoise Tisseur, Nick Trefethen, Joan Walsh, Barry White.

Working with the SIAM staff was once again a pleasure. I thank, in particular, my copy editor Beth Gallagher, Vickie Kearn and Mary Rose Muccie.

This book was typeset in LaTeX using the **book** document style and various LaTeX packages. The references were prepared in BibTeX and the index with MakeIndex. I used software from the emTeX distribution, running on a Pentium workstation. I used text editors The Semware Editor (Semware Corporation) and GNU Emacs (Free Software Foundation) and checked spelling with PC-Write (Quicksoft).

Manchester   Nicholas J. Higham
January 1998

Preface to the First Edition

In this book I aim to describe most of what a scientist needs to know about mathematical writing. Although the emphasis is strongly on *mathematical* writing, many of the points and issues I discuss are relevant to *scientific* writing in general. My main target audience is graduate students. They often have little experience or knowledge of technical writing and are daunted by the task of writing a report or thesis. The advice given here reflects what I have learned in the ten years since I wrote my first research report as a graduate student and describes what I would have liked to know as I started to write that first report. I hope that as well as being a valuable resource for graduate students, this book will also be of use to practising scientists.

The book has grown out of notes for a short lecture course on mathematical writing that I gave at the University of Manchester in May 1992. As I prepared the course I realized that, although several excellent articles and books on mathematical writing are available (notably those by Halmos (1970) [121], Gillman (1987) [104] and Knuth, Larrabee and Roberts (1989) [164]), none functions as a comprehensive handbook that can be both read sequentially and used as a reference when questions about mathematical writing arise. I have attempted to provide such a handbook. (I hope that the comment of one journal referee, "This paper fills a much needed gap in the literature", is not applicable to this book!)

As well as covering standard topics such as English usage, the anatomy of a research paper, and revising a draft, I examine in detail four topics that are usually discussed only briefly, if at all, in books on technical writing.

- The whole publication process, from submission of a manuscript to its appearance as a paper in a journal.

- Writing when English is a foreign language.

- How to write slides for a talk.

- The use of computers in writing and research. In particular, I discuss modern practices such as computerized typesetting in TeX, the use of computer tools for indexing and checking spelling and style, and electronic mail and ftp (file transfer protocol).

An important feature of the book is that many examples are given to illustrate the ideas and principles discussed. In particular, Chapter 7 contains a collection of extracts from the mathematics and computer science literature, with detailed comments on how each extract can be improved.

In writing the book I have been helped and influenced by many people. Several people read the entire manuscript at one or more of its various stages, offered constructive suggestions, encouragement and advice, and made sure I said what I meant and meant what I said. They are

Ian Gladwell, Des Higham, Doris Higham, Nil Mackey, Fred Schneider, Pete (G. W.) Stewart and Nick Trefethen.

Other people who have read portions of the book and have given help, suggestions or advice are

Carl de Boor, David Carlisle, Valérie Fraysse, Paul Halmos, Bo Kågström, Philip Knight, Sven Leyffer, Steve Mackey, Volker Mehrmann, June O'Brien, Pythagoras Papadimitriou, Beresford Parlett, Nigel Ray, Stephan Rudolfer, Bob Sandling, Zdenek Strakos, Gil Strang, Charlie Van Loan, Rossana Vermiglio, Joan Walsh, Barry White, and Yuanjing Xu.

I thank all these people, together with many others who have answered my questions and made suggestions. In researching the contents of the book I was inspired by many of the references listed in the bibliography, and learned a lot from them. I acknowledge the Nuffield Foundation for the support of a Nuffield Science Research Fellowship, during the tenure of which I wrote much of the book. Last, but not least, I thank the SIAM staff for their help and advice in the production of the book—in particular, Susan Ciambrano, Beth Gallagher and Tricia Manning. I would be happy to receive comments, notification of errors, and suggestions for improvement, which I will collect for inclusion in a possible second edition.

This book was typeset in LATEX using the book document style with G. W. Stewart's jeep style option. I prepared the references with BiBTEX and the index with MakeIndex. I used text editors Qedit (Semware) and GNU Emacs (Free Software Foundation), and checked spelling with PC-Write (Quicksoft).

Manchester   Nicholas J. Higham
December 1992

Chapter 1
General Principles

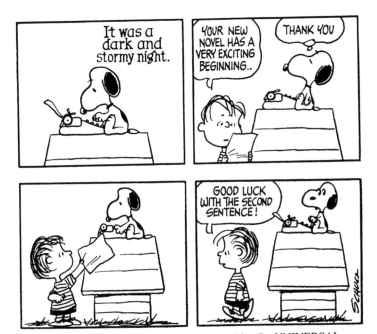

Good writing ... is clear thinking made visible.
— AMBROSE BIERCE, *Write it Right:*
A Little Blacklist of Literary Faults (1937)

A writer needs three qualities:
creativity, originality, clarity and a good short term memory.
— DESMOND J. HIGHAM, *More Commandments of Good Writing* (1992)

1

Writing helps you to learn. Writing is not simply a task to be done once research or other preparation is completed—it can be an integral part of the work process. Baker explains it well [13]:

> In writing, you clarify your own thoughts, and strengthen your conviction. Indeed, you probably grasp your thoughts for the first time. Writing is a way of thinking. Writing actually creates thought, and generates your ability to think: you discover thoughts you hardly knew you had, and come to know what you know. You learn as you write.

Writing brings out gaps in your understanding, by forcing you to focus on steps in your thinking that you might otherwise skip. (In this respect, writing is similar to explaining your ideas to a colleague.) Writing down partial research results as your work progresses can reveal structure and suggest further areas of development. Zinsser has written a delightful book that explores the idea of writing to learn [303].

Good writing reflects clear thinking. It is very hard for a woolly thinker to produce clear prose. Clear thinking leads to good organization, a vital ingredient of technical writing. A well-organized piece is much easier to write than a badly organized one. If you find a particular piece difficult to write it may be because you have not found the right structure to express your ideas.

Writing is difficult! Zinsser [304] says "It's one of the hardest things that people do." It is often difficult to get started. One solution, which works for some, is to force yourself to write something, however clumsy it may be, for it is often easier to modify something you have written previously than to compose from scratch.

The most fundamental tenet of technical writing is to keep your prose simple and direct. Much of written English is unnecessarily complicated. In writing up your research you are aiming at a relatively small audience, so it is important not to alienate part of it with long-winded or imprecise text. English may not be the first language of many of your readers—they, particularly, will appreciate plain writing. Aim for economy of words. Early drafts can usually be reduced in length substantially with consequent improvements in readability (see Chapter 7).

Probably the best way to improve your writing skills is to receive, and learn from, constructive criticism. Ask a colleague to read and comment on your writing. Another reader can often find errors and suggest improvements that you miss because of your familiarity with the work. Criticism can be difficult to take, but it should be welcomed; it is a privilege to have someone else take the time to comment on your writing.

Another way to improve your writing is to read as much as you can, always with a critical eye. In particular, I recommend perusal of some of the following mathematical books. They are by excellent writers, each of whom has his own distinctive style (this selection is inevitably biased towards my own area of research):

▷ Forman S. Acton (1970), *Numerical Methods That Work* [3].

▷ Albert H. Beiler (1966), *Recreations in the Theory of Numbers* [19].

▷ David M. Burton (1980), *Elementary Number Theory* [44].

▷ Gene H. Golub and Charles F. Van Loan (1996), *Matrix Computations* [108].

▷ Paul R. Halmos (1982), *A Hilbert Space Problem Book* [125].

▷ Donald E. Knuth (1973–1981), *The Art of Computer Programming* [157]. (Knuth was awarded the 1986 Leroy P. Steele Prize by the AMS for these three volumes.)

▷ Beresford N. Parlett (1998), *The Symmetric Eigenvalue Problem* [217].

▷ G. W. Stewart (1973), *Introduction to Matrix Computations* [261].

▷ Gilbert Strang (1986), *Introduction to Applied Mathematics* [262].

Also worth studying are papers or books that have won prizes for expository writing in mathematics. Appendix E lists winners of the Chauvenet Prize, the Lester R. Ford Award, the George Polya Award, the Carl B. Allendoerfer Award, the Beckenbach Book Prize and the Merten M. Hasse Prize.

Chapter 2
Writer's Tools and Recommended Reading

I use three dictionaries almost every day.
— JAMES A. MICHENER, *Writer's Handbook* (1992)

The purpose of an ordinary dictionary is simply
to explain the meaning of the words . . . .
The object aimed at in the present undertaking is exactly the converse of this:
namely,—The idea being given, to find the word, or words,
by which that idea may be most fitly and aptly expressed.
— PETER MARK ROGET, *Thesaurus of English Words and Phrases* (1852)

The dictionary and thesaurus interruptions are usually
not about meaning in the gross sense
(what's the correct use of "oppugn"),
but about precision, and about finding the right word. . .
What did the examples that von Neumann and I constructed
do to the conjugacy conjecture for shifts. . .
did they contradict, contravene, gainsay, dispute,
disaffirm, disallow, abnegate, or repudiate it?. . .
Writing can stop for 10 or 15 minutes while I search and weigh.
— PAUL R. HALMOS, *I Want to be a Mathematician:*
An Automathography in Three Parts (1985)

Mathematicean. *One that is skilled in Augurie, Geometrie, and Astronomie.*
— HENRY COCKERAM[1], *English Dictionarie* (1623)

---

[1]Quoted in [255].

5

2.1. Dictionaries and Thesauruses

Apart from pen, paper and keyboard, the most valuable tool for a writer in any subject is a dictionary. Writing or reading in the mathematical sciences you will come across questions such as the following:

1. What is the plural of modulus: moduli or moduluses?

2. Which of parameterize and parametrize is the correct spelling?

3. What is a gigaflop?

4. When was the mathematician Abel born and what was his nationality?

5. What is the meaning of mutatis mutandis?

6. Who was Procrustes (as in the "orthogonal Procrustes problem")?

7. When should you use "special" and when "especial"?

8. What are the differences between mind-bending, mind-blowing and mind-boggling?

All the answers can be found in general-purpose dictionaries (and are given at the end of this chapter). As these questions illustrate, dictionaries are invaluable for choosing a word with just the right shade of meaning, checking on spelling and usage, and even finding encyclopedic information. Furthermore, the information about a word's history provided in a dictionary etymology can make it easier to use the word precisely.

The most authoritative dictionary is the *Oxford English Dictionary* (OED) [215]. It was originally published in parts between 1884 and 1928, and a four volume supplement was produced from 1972–1986. A twenty volume second edition of the dictionary was published in 1989; it defines more than half a million words, using 2.4 million illustrative quotations. The OED traces the history of words from around 1150. In 1992 a compact disc (CD-ROM) version of the OED was published. It contains the full text of the printed version (at about a third of the price) and the accompanying software includes powerful search facilities. Other large dictionaries are *Webster's Third New International Dictionary* [294], which was published in the United States in 1961 and has had three supplements, *The American Heritage Dictionary of the English Language* [7], the *Random House Unabridged Dictionary* [233], and *The New Shorter Oxford English Dictionary*, in two volumes [214].

For everyday use the large dictionaries are too unwieldy and too thorough, so a more concise dictionary is needed. *The Concise Oxford Dictionary* (COD) [213] is now in its ninth edition (1995). It is the favourite of many, and is suitable for American use, as American spellings and usages are included. (The COD was my main dictionary of reference in writing this book.) Other dictionaries suitable for regular use by the writer include, from the United States:

- *The American Heritage College Dictionary* [6].

- *The Random House Webster's College Dictionary* [234].

- *Merriam-Webster's Collegiate Dictionary* [203]. Most main entries state the date of first recorded use of the word. Contains usage and synonym notes and appendices "Biographical Names" and "Geographical Names".

- *Webster's New World College Dictionary* [293].

From Britain:

- *The Chambers Dictionary* [54]. Renowned for its rich vocabulary, which includes literary terms, Scottish words and many archaic and obsolete words. Also contains some humorous entries: *éclair* is defined as "a cake, long in shape but short in duration ...".

- The *Collins English Dictionary* [60]. Contains extensive encyclopedic entries, both biographical and geographical, strong coverage of scientific and technical vocabulary, and usage notes.

- The *Longman Dictionary of the English Language* [182]. The same comments apply as for the Collins. Has an extensive collection of notes on usage, synonyms and word history.

The American dictionaries listed, but not the British ones, show allowable places to divide words when they must be broken and hyphenated at the end of a line.

To make good use of dictionaries, it helps to be aware of some of their characteristics.

Order of definitions. For words with several meanings, most dictionaries give the most common or current meanings first, but some give meanings in their historical sequence. The historical order is the one used by the *Oxford English Dictionary*, since its purpose is to trace the development of words from their first use to the present day. The *Merriam-Webster's Collegiate* also uses the historical order, but for a desk dictionary intended for quick reference this order can be disorienting. For example, under the headword *nice*, *Merriam-Webster's Collegiate* lists "showing fastidious or finicky tastes" before "pleasing, agreeable".

Etymologies. Etymologies vary in their location within an entry, in the style in which they are presented (for example, the symbol < may be used for "from"), and in their depth and amount of detail. Some words with interesting etymologies are *diploma*, *OK*, *shambles*, *symposium*, and *sine*.

Scientific and technical vocabulary. Since there are vastly more scientific and technical terms than any general dictionary can accommodate, there is much variation in the coverage provided by different dictionaries.

Up-to-date vocabulary. The constantly changing English language is monitored by lexicographers (Johnson's "harmless drudges"), who add new words and meanings to each new edition of their dictionaries. Coverage of modern vocabulary varies between dictionaries, depending on the year of publication and the compilers' tastes and citation files (which usually include material submitted by the general public).

British versus American spelling and usage. Since much mathematical science is written for an international audience it is useful to be able to check differences in British and American spelling and usage. Most British and American dictionaries are good in this respect.

General-purpose dictionaries do not always give correct definitions of mathematical terms. In a comparison of eight major British and American dictionaries I found errors in definitions of terms such as *determinant*, *eigenvector*[2], *polynomial*, and *power series* [141].

Annotated lists of dictionaries and usage guides are given by Stainton [253], [254]. Comparisons and analyses of dictionaries are also given by Quirk and Stein [232, Chap. 11] and Burchfield [43].

Specialized dictionaries can also be useful to the mathematical writer. There are many dictionaries of mathematics, one example being the Penguin dictionary [206], which is small and inexpensive yet surprisingly thorough. Schwartzman's *The Words of Mathematics* [247] explains the etymology of words used in mathematics (see also [248]).

The synonyms provided in a thesaurus can be helpful in your search for an elusive word or a word with the right connotation. *Roget's Thesaurus*, first published in 1852, is the classic one. The words in *Roget's Thesaurus* are traditionally arranged according to the ideas they express, instead of alphabetically, though versions are now available in dictionary form. The *Bloomsbury Thesaurus* [32] is arranged according to a new classification

---

[2]One dictionary offers this definition of eigenvector: a vector that in one dimension under a given rotational, reflectional, expanding, or shrinking operation becomes a number that is a multiple of itself.

designed to be more appropriate for modern English than that of Roget, and it has a very detailed index. Rodale's *The Synonym Finder* [236] is a large thesaurus arranged alphabetically. Thesauruses are produced by all the major publishers of dictionaries.

2.2. Usage and Style Guides

Every writer should own and read a guide to English usage. One of the most accessible is *The Elements of Style* by Strunk and White [263]. Zinsser [304] says this is "a book that every writer should read at least once a year", and, as if following this advice, Luey [185] says "I read it once a year without fail." An even shorter, but equally readable, guide is Lambuth et al.'s *The Golden Book on Writing* [170]. Fowler's *Dictionary of Modern English Usage* [83] is a much longer and more detailed work, as is its predecessor, *The King's English,* by the Fowler brothers [84]. A favourite of mine is the revision [298] by Flavell and Flavell of the 1962 *Current English Usage* by Wood. Gowers's influential *Complete Plain Words* [115] stems from his *Plain Words* of 1948, which was written to improve the use of English in the British civil service. Partridge's *Usage and Abusage* [218] is another valuable guide, this one in dictionary form.

Excellent advice on punctuation is given by Carey in *Mind the Stop* [52] and by Bernstein [28]. For a whimsical treatment, see *The New Well-Tempered Sentence* by Gordon [112].

Bryson's *Dictionary of Troublesome Words* [41] offers practical, witty advice on usage, while Safire [243] presents fifty "fumblerules" (mistakes that call attention to the rules) accompanied by pithy explanatory essays. The books *On Newspaper Style* and *English our English* by Waterhouse [287], [288] make fascinating and informative reading, though they are hard to use for reference since they lack an index; [287] is a standard handbook for journalists, but is of much wider interest. Baker's *The Practical Stylist* [13] is a widely used course text on writing; it has thorough discussions of usage, style and revision and gives many illustrative examples. Day's *Scientific English* [69] contains general advice on grammar and usage, with particular reference to English in scientific writing. Perry's *The Fine Art of Technical Writing* [221] offers selective, practical advice on the psychology, artistry and technique of technical writing, which the author defines as "all writing other than fiction". In *Miss Thistlebottom's Hobgoblins* [26] Bernstein provides an antidote for those brainwashed by over-prescriptive usage guides, in the form of letters to his (fictional) English schoolteacher. Two other books by Bernstein, *The Careful Writer* [25] and *Dos, Don'ts and Maybes of English Usage* [27], are also useful guides. Gordon's *The Transitive Vampire* [111] is a grammar guide in the same fanciful vein as [112].

The Chicago Manual of Style [58], first published in 1906, is a long and comprehensive guide to book production, style and printing. It is the standard reference for authors and editors in many organizations. It includes chapters on typesetting mathematics and preparing bibliographies and indexes. Turabian's *A Manual for Writers of Term Papers, Theses, and Dissertations* [278], first published in 1937, is based on the guidelines in *The Chicago Manual of Style* but its aim is more limited, as defined in the title, so it does not discuss bookmaking and copy editing. *Words into Type* [249] is another thorough guide for authors and editors, covering manuscript and index preparation, copy editing style, grammar, typographical style and the printing process. Other valuable references on editing, copy editing and proofreading are *Hart's Rules* [131], which describes the house style of Oxford University Press; Butcher's *Copy-Editing* [45], which is regarded as the standard British work on copy editing; Eisenberg's *Guide to Technical Editing* [77]; O'Connor's *How to Copyedit Scientific Books and Journals* [208]; Stainton's *The Fine Art of Copyediting* [254]; and Tarutz's *Technical Editing* [270].

Some interesting techniques for revising a sentence by analysing its structure are presented by Lanham in *Revising Prose* [175].

2.3. Technical Writing Guides

Several guides to mathematical writing are available. Halmos's essay "How to Write Mathematics" [121] is essential reading for every mathematician; it contains much sound advice not found elsewhere. Halmos's "automathography" [127] includes insight into mathematical writing, editing and refereeing; it begins with the sentence "I like words more than numbers, and I always did." Transcripts of a lecture course called "Mathematical Writing" that was given by Knuth in 1987 at Stanford are collected in *Mathematical Writing* [164], which I highly recommend. This manual contains many anecdotes and insights related by Knuth and his guest lecturers, including Knuth's battle with the copy editors at *Scientific American* and his experiences in writing the book *Concrete Mathematics* [116]. Other very useful guides are Flanders's article [80] for authors who write in the journal *American Mathematical Monthly*; Gillman's booklet *Writing Mathematics Well* [104] on preparing manuscripts for Mathematical Association of America journals; Steenrod's essay "How to Write Mathematics" [256]; Krantz's wide-ranging *A Primer of Mathematical Writing* [167]; and Swanson's guide *Mathematics into Type* [267] for mathematical copy editors and authors. Knuth's book on TeX [161] contains much general advice on how to typeset mathematics, and an old guide to this subject is *The Printing of Mathematics* [55].

Most books and papers on mathematical writing, including this one, are aimed primarily at graduate students and advanced undergraduate students. Maurer [197] gives advice on mathematical writing aimed specifically at undergraduate students, covering a number of basic issues omitted elsewhere.

Guides to writing in other scientific disciplines often contain much that is relevant to the mathematical writer; an example is the book by Pechenik [219], which is aimed at biology students. General guides to scientific writing that I recommend are those by Barrass [14], [15], Cooper [62], Ebel, Bliefert and Russey [76], Kirkman [153], O'Connor [209] (this is a revised and extended version of an earlier book by O'Connor and Woodford [210]), and Turk and Kirkman [280]. The book edited by Woodford [300] contains three examples of short papers in both original and revised forms, with detailed annotations. Particularly informative and pleasant to read are Booth's *Communicating in Science* [36] and Day's *How to Write and Publish a Scientific Paper* [68].

The journal *IEEE Transactions on Professional Communication* publishes papers on many aspects of technical communication, including how to write papers and give talks. A selection of 63 papers from this and other journals is collected in *Writing and Speaking in the Technology Professions: A Practical Guide* [18].

How to Do It [180] contains 47 chapters that give advice for medical doctors, but many of them are of general interest to scientists. Chapter titles include "Write a Paper", "Referee a Paper", "Attract the Reader", "Review a Book", "Use an Overhead Projector", and "Apply for a Research Grant". Many of the chapters originally appeared in the *British Medical Journal.*

Van Leunen's *A Handbook for Scholars* [283] is a unique and indispensable guide to the mechanics of scholarly writing, covering reference lists, quotations, citations, footnotes and style. This is the place to look if you want to know how to prepare a difficult reference or quotation (what date to list for a reprint of a work from a previous century, or how to punctuate a quotation placed in mid-sentence). There is also an appendix on how to prepare a CV. Luey's *Handbook for Academic Authors* [185] offers much useful advice to the writer of an academic book.

O'Connor has written a book about how to edit and manage scientific books and journals [207].

Thirty-one essays discussing how writing is being used to teach mathematics in undergraduate courses are contained in *Using Writing to Teach Mathematics* [259].

A useful source for examples of expository mathematical writing is the annotated bibliography of Gaffney and Steen [87], which contains more than 1100 entries.

Finally, Pemberton's book *How to Find Out in Mathematics* [220] tells you precisely what the title suggests. It includes information on mathematical dictionaries (including interlingual ones) and encyclopedias, mathematical histories and biographies, and mathematical societies, periodicals and abstracts. Although it appeared in 1969, the book is still worth consulting.

2.4. General Reading

The three books by Zinsser [302], [303], [304] are highly recommended; all are informative and beautifully written. In *Writing with a Word Processor* [302] Zinsser summarizes his experience in moving to a computer from his trusty typewriter. His book *Writing to Learn* contains chapters on "Writing Mathematics" and "Writing Physics and Chemistry"; they explain how writing can be used in the teaching of these subjects and give examples of good writing. Michener's *Writer's Handbook* [204] provides insight into how this prodigious writer worked. The reader is led through the development of parts of two of Michener's books (one fiction, one nonfiction), from early drafts to proofs to the published versions.

Mitchell [205] gives hints on writing with a computer, with good examples of how to revise drafts.

Valuable insight into the English language—its history, its eccentricities, and its uses—is provided by Bryson [42], Crystal [66] and Potter [229].

Answers to the Questions at the Start of the Chapter

1. The plural of modulus is *moduli*.

2. The *Concise Oxford Dictionary* gives only the spelling parametrize, but the *Longman Dictionary of the English Language*, *Merriam-Webster's Collegiate Dictionary* and *Oxford English Dictionary* give both parameterize and parametrize.

3. From the *Collins English Dictionary*: "**gigaflop**... *n. Computer technol.* a measure of processing speed, consisting of a thousand million floating-point operations a second. [C20 ...]".

4. From the entry for *Abelian group* in the *Collins English Dictionary*: "Niels Henrik *Abel* (1802–29), Norwegian mathematician".

5. Mutatis mutandis means "with necessary changes" (*The Chambers Dictionary*).

6. Procrustes was "a villainous son of Poseidon in Greek mythology who forces travelers to fit into his bed by stretching their bodies or cutting off their legs" (*Merriam-Webster's Collegiate Dictionary*).

7. From the *Collins English Dictionary* (usage note after *especial*): "*Especial* and *especially* have a more limited use than *special* and *specially*. *Special* is always used in preference to *especial* when the sense is one of being out of the ordinary ... Where an idea of pre-eminence or individuality is involved, either *especial* or *special* may be used."

8. From the *Longman Dictionary of the English Language*, all three words being labelled *adj, informal*:

> *mind-bending* means "at the limits of understanding or credibility",
> *mind-blowing* means "**1** of or causing a psychic state similar to that produced by a psychedelic drug **2** mentally or emotionally exhilarating; overwhelming",
> *mind-boggling* means "causing great surprise or wonder".

Chapter 3
Mathematical Writing

Suppose you want to teach the "cat" concept to a very young child.
Do you explain that a cat is a relatively small,
primarily carnivorous mammal with retractile claws,
a distinctive sonic output, etc.?
I'll bet not.
You probably show the kid a lot of different cats,
saying "kitty" each time, until it gets the idea.
To put it more generally,
generalizations are best made by abstraction from experience.
— RALPH P. BOAS, *Can We Make Mathematics Intelligible?* (1981)

A good notation should be unambiguous, pregnant, easy to remember;
it should avoid harmful second meanings, and
take advantage of useful second meanings;
the order and connection of signs should
suggest the order and connection of things.
— GEORGE POLYA, *How to Solve It* (1957)

We have not succeeded in finding or constructing
a definition which starts out
"A Bravais lattice is ...";
the sources we have looked at say
"That was a Bravais lattice."
— CHARLES KITTEL, *Introduction to Solid State Physics* (1971)

Notation is everything.
— CHARLES F. VAN LOAN, *FFTs and the Sparse Factorization Idea* (1992)

The mathematical writer needs to be aware of a number of matters specific to mathematical writing, ranging from general issues, such as choice of notation, to particular details, such as how to punctuate mathematical expressions. In this chapter I begin by discussing some of the general issues and then move on to specifics.

3.1. What Is a Theorem?

What are the differences between theorems, lemmas, and propositions? To some extent, the answer depends on the context in which a result appears. Generally, a *theorem* is a major result that is of independent interest. The proof of a theorem is usually nontrivial. A *lemma*[3] is an auxiliary result—a stepping stone towards a theorem. Its proof may be easy or difficult. A straightforward and independent result that is worth encapsulating but that does not merit the title of a theorem may also be called a lemma. Indeed, there are some famous lemmas, such as the Riemann–Lebesgue Lemma in the theory of Fourier series and Farkas's Lemma in the theory of constrained optimization. Whether a result should be stated formally as a lemma or simply mentioned in the text depends on the level at which you are writing. In a research paper in linear algebra it would be inappropriate to give a lemma stating that the eigenvalues of a symmetric positive definite matrix are positive, as this standard result is so well known; but in a textbook for undergraduates it would be sensible to formalize this result.

It is not advisable to label all your results theorems, because if you do so you miss the opportunity to emphasize the logical structure of your work and to direct attention to the most important results. If you are in doubt about whether to call a result a lemma or a theorem, call it a lemma.

The term *proposition* is less widely used than lemma and theorem and its meaning is less clear. It tends to be used as a way to denote a minor theorem. Lecturers and textbook authors might feel that the modest tone of its name makes a proposition appear less daunting to students than a theorem. However, a proposition is not, as one student thought, "a theorem that might not be true".

A *corollary* is a direct or easy consequence of a lemma, theorem or proposition. It is important to distinguish between a corollary, which does not imply the parent result from which it came, and an extension or generalization of a result. Be careful not to over-glorify a corollary by failing to label it as such, for this gives it false prominence and obscures the role of the parent result.

---

[3]The plural of lemma is lemmata, or, more commonly, lemmas.

How many results are formally stated as lemmas, theorems, propositions or corollaries is a matter of personal style. Some authors develop their ideas in a sequence of results and proofs interspersed with definitions and comments. At the other extreme, some authors state very few results formally. A good example of the latter style is the classic book *The Algebraic Eigenvalue Problem* [296] by Wilkinson, in which only four titled theorems are given in 662 pages. As Boas [33] notes, "A great deal can be accomplished with arguments that fall short of being formal proofs."

A fifth kind of statement used in mathematical writing is a *conjecture*—a statement that the author thinks may be true but has been unable to prove or disprove. The author will usually have some strong evidence for the veracity of the statement. A famous example of a conjecture is the Goldbach conjecture (1742), which states that every even number greater than 2 is the sum of two primes; this is still unproved. One computer scientist (let us call him Alpha) joked in a talk "This is the Alpha and Beta conjecture. If it turns out to be false I would like it to be known as Beta's conjecture." However, it is not necessarily a bad thing to make a conjecture that is later disproved: identifying the question that the conjecture aims to answer can be an important contribution.

A *hypothesis* is a statement that is taken as a basis for further reasoning, usually in a proof—for example, an induction hypothesis. Hypotheses that stand on their own are uncommon; two examples are the Riemann hypothesis and the continuum hypothesis.

3.2. Proofs

Readers are often not very interested in the details of a proof but want to know the outline and the key ideas. They hope to learn a technique or principle that can be applied in other situations. When readers do want to study the proof in detail they naturally want to understand it with the minimum of effort. To help readers in both circumstances, it is important to emphasize the structure of a proof, the ease or difficulty of each step, and the key ideas that make it work. Here are some examples of the sorts of phrases that can be used (most of these are culled from proofs by Parlett in [217]).

> The aim/idea is to
> Our first goal is to show that
> Now for the harder part.
> The trick of the proof is to find
> ... is the key relation.
> The only, but crucial use of ... is that

To obtain ... a little manipulation is needed.
The essential observation is that

When you omit part of a proof it is best to indicate the nature and length of the omission, via phrases such as the following.

It is easy/simple/straightforward to show that
Some tedious manipulation yields
An easy/obvious induction gives
After two applications of ... we find
An argument similar to the one used in ... shows that

You should also strive to keep the reader informed of where you are in the proof and what remains to be done. Useful phrases include

First, we establish that
Our task is now to
Our problem reduces to
It remains to show that
We are almost ready to invoke
We are now in a position to
Finally, we have to show that

The end of a proof is often marked by the halmos symbol □ (see the quote on page 24). Sometimes the abbreviation QED (Latin: quod erat demonstrandum = which was to be demonstrated) is used instead.

There is much more to be said about writing (and devising) proofs. References include Franklin and Daoud [85], Garnier and Taylor [101], Lamport [173], Leron [177] and Polya [228].

3.3. The Role of Examples

A pedagogical tactic that is applicable to all forms of technical writing (from teaching to research) is to discuss specific examples before the general case. It is tempting, particularly for mathematicians, to adopt the opposite approach, but beginning with examples is often the more effective way to explain (see Boas's article [33] and the quote from it at the beginning of this chapter, a quote that itself illustrates this principle!).

A good example of how to begin with a specific case is provided by Strang in Chapter 1 of *Introduction to Applied Mathematics* [262]:

The simplest model in applied mathematics is a system of linear equations. It is also by far the most important, and we begin this book with an extremely modest example:

$$2x_1 + 4x_2 = 2,$$
$$4x_1 + 11x_2 = 1.$$

After some further introductory remarks, Strang goes on to study in detail both this 2×2 system and a particular 4×4 system. General $n \times n$ matrices appear only several pages later.

Another example is provided by Watkins's *Fundamentals of Matrix Computations* [289]. Whereas most linear algebra textbooks introduce Gaussian elimination for general matrices before discussing Cholesky factorization for symmetric positive definite matrices, Watkins reverses the order, giving the more specific but algorithmically more straightforward method first.

An exercise in a textbook is a form of example. I saw a telling criticism in one book review that complained "The first exercise in the book was pointless, so why do the others?" To avoid such criticism, it is important to choose exercises and examples that have a clear purpose and illustrate a point. The first few exercises and examples should be among the best, to gain the reader's confidence. The same reviewer complained of another book that "it hides information in exercises and contains exercises that are too difficult." Whether such criticism is valid depends on your opinion of what are the key issues to be transmitted to the reader and on the level of the readership. Again, it helps to bear such potential criticism in mind when you write.

3.4. Definitions

Three questions to be considered when formulating a definition are "why?", "where?" and "how?" First, ask yourself why you are making a definition: is it really necessary? Inappropriate definitions can complicate a presentation and too many can overwhelm a reader, so it is wise to imagine yourself being charged a large sum for each one. Instead of defining a square matrix A to be contractive with respect to a norm $\| \cdot \|$ if $\|A\| < 1$, which is not a standard definition, you could simply say "A with $\|A\| < 1$" whenever necessary. This is easy to do if the property is needed on only a few occasions, and saves the reader having to remember what "contractive" means. For notation that is standard in a given subject area, judgement is needed to decide whether the definition should be given. Potential confusion can often be avoided by using redundant words. For example, if $\rho(A)$ is not obviously the spectral radius of the matrix A you can say "the spectral radius $\rho(A)$".

The second question is "where?" The practice of giving a long sequence of definitions at the start of a work is not recommended. Ideally, a definition should be given in the place where the term being defined is first used. If it is given much earlier, the reader will have to refer back, with a possible loss of concentration (or worse, interest). Try to minimize the distance between a definition and its place of first use.

It is not uncommon for an author to forget to define a new term on its first occurrence. For example, Steenrod uses the term "grasshopper reader" on page 6 of his essay on mathematical writing [256], but does not define it until it occurs again on the next page.

To reinforce notation that has not been used for a few pages you may be able to use redundancy. For example, "The optimal steplength α^* can be found as follows." This implicit redefinition either reminds readers what α^* is, or reassures them that they have remembered it correctly.

Finally, how should a term be defined? There may be a unique definition or there may be several possibilities (a good example is the term M-matrix, which can be defined in at least fifty different ways [23]). You should aim for a definition that is short, expressed in terms of a fundamental property or idea, and consistent with related definitions. As an example, the standard definition of a normal matrix is a matrix $A \in \mathbb{C}^{n \times n}$ for which $A^*A = AA^*$ (where $*$ denotes the conjugate transpose). There are at least 70 different ways of characterizing normality [119], but none has the simplicity and ease of use of the condition $A^*A = AA^*$.

By convention, *if* means *if and only if* in definitions, so do not write "The graph G is connected if and only if there is a path from every node in G to every other node in G." Write "The graph G is connected if there is a path from every node in G to every other node in G" (and note that this definition can be rewritten to omit the symbol G). It is common practice to italicize the word that is being defined: "A graph is *connected* if there is a path from every node to every other node." This has the advantage of making it perfectly clear that a definition is being given, and not a result. This emphasis can also be imparted by writing "A graph is defined to be connected if ... ", or "A graph is said to be connected if ...."

If you have not done so before, it is instructive to study the definitions in a good dictionary. They display many of the attributes of a good mathematical definition: they are concise, precise, consistent with other definitions, and easy to understand.

Definitions of symbols are usually made with a simple equality, perhaps preceded by the word "let" if they are in-line, as in "let $q(x) = ax^2 + bx + c$." Various other notations have been devised to give emphasis to a definition, including

$$q(x) := ax^2 + bx + c,$$
$$ax^2 + bx + c =: q(x),$$
$$q(x) \stackrel{\text{def}}{=} ax^2 + bx + c,$$
$$q(x) \equiv ax^2 + bx + c,$$
$$q(x) \triangleq ax^2 + bx + c.$$

If you use one of these special notations you must use it consistently, otherwise the reader may not know whether a straightforward equality is meant to be a definition.

3.5. Notation

Consider the following extract.

> Let $\widehat{H}_k = Q_k^H \widetilde{H}_k Q_k$, partition $X = [X_1, \ X_2]$ and let $\mathcal{X} =$ range(X_1). Let U^* denote the nearest orthonormal matrix to X_1 in the 2-norm.

These two sentences are full of potentially confusing notation. The distinction between the hat and the tilde in $\widehat{H}_k$ and $\widetilde{H}_k$ is slight enough to make these symbols difficult to distinguish. The symbols $\mathcal{X}$ and X are also too similar for easy recognition. Given that $\mathcal{X}$ is used, it would be more consistent to give it a subscript 1. The name H_k is unfortunate, because H is being used to denote the conjugate transpose, and it might be necessary to refer to $\widetilde{H}_k^H$! Since A^* is a standard synonym for A^H, the use of a superscripted asterisk to denote optimality is confusing.

As this example shows, the choice of notation deserves careful thought. Good notation strikes a balance among the possibly conflicting aims of being readable, natural, conventional, concise, logical and aesthetically pleasing. As with definitions, the amount of notation should be minimized.

Although there are 26 letters in the alphabet and nearly as many again in the Greek alphabet, our choice diminishes rapidly when we consider existing connotations. Traditionally, ϵ and δ denote small quantities, i, j, k, m and n are integers (or i or j the imaginary unit), λ is an eigenvalue and π and e are fundamental constants; π is also used to denote a permutation. These conventions should be respected. But by modifying and combining eligible letters we widen our choice. Thus γ and A yield, for example, $\widehat{A}$, $\overline{A}$, $\widetilde{A}$, A', γ_A, A_γ, $\mathbf{A}$, $\mathcal{A}$, $\mathbb{A}$.

Particular areas of mathematics have their own notational conventions. For example, in numerical linear algebra lower case Greek letters represent scalars, lower case roman letters represent column vectors, and upper case Greek or roman letters represent matrices. This convention was introduced by Householder [143].

In his book on the symmetric eigenvalue problem [217], Parlett uses the symmetric letters A, H, M, T, U, V, W, X, Y to denote symmetric matrices and the symmetric Greek letters Λ, Θ, Φ, Δ to denote diagonal matrices. Actually, the roman letters printed above are not symmetric because they are slanted, but Parlett's book uses a sans serif mathematics font that yields the desired symmetry. Parlett uses this elegant, but restrictive, convention to good effect.

We can sometimes simplify an expression by giving a meaning to extreme cases of notation. Consider the display

$$\beta_{ij} = \begin{cases} 0, & i > j, \\ \dfrac{1}{u_j}, & i = j, \\ \dfrac{1}{u_j} \displaystyle\prod_{r=i}^{j-1} \left(\dfrac{-c_r}{u_r} \right), & i < j. \end{cases}$$

There are really only two cases: $i > j$ and $i \leq j$. This structure is reflected and the display made more compact if we define the empty product to be 1, and write

$$\beta_{ij} = \begin{cases} 0, & \text{if } i > j, \\ \dfrac{1}{u_j} \displaystyle\prod_{r=i}^{j-1} \left(\dfrac{-c_r}{u_r} \right), & \text{if } i \leq j. \end{cases}$$

(Here, I have put "if" before each condition, which is optional in this type of display.) Incidentally, note that in a matrix product the order of evaluation needs to be specified: $\prod_{i=1}^{n} A_i$ could mean $A_1 A_2 \ldots A_n$ or $A_n A_{n-1} \ldots A_1$.

Notation also plays a higher level role in affecting the way a method or proof is presented. For example, the $n \times n$ matrix multiplication $C = AB$ can be expressed in terms of scalars,

$$c_{ij} = \sum_{k=1}^{n} a_{ik} b_{kj}, \qquad 1 \leq i, j \leq n,$$

or at the matrix-vector level,

$$C = [\, Ab_1, \quad Ab_2, \quad \ldots, \quad Ab_n \,],$$

where $B = [b_1, b_2, \ldots, b_n]$ is a partition into columns. One of these two viewpoints may be superior, depending on the circumstances. A deeper example is provided by the fast Fourier transform (FFT). The discrete Fourier transform (DFT) is a product $y = F_n x$, where F_n is the unitary Vandermonde matrix with (r, s) element $\omega^{(r-1)(s-1)}$ $(1 \leq r, s \leq n)$, and $\omega = \exp(-2\pi i/n)$. The FFT is a way of forming this product in $O(n \log n)$ operations. It is traditionally expressed through equations such as the following (copied from a numerical methods textbook):

$$\sum_{j=0}^{n-1} e^{2\pi i j k/n} f_j = \sum_{j=0}^{n/2-1} e^{2\pi i k j/(n/2)} f_{2j} + \omega^k \sum_{j=0}^{n/2-1} e^{2\pi i k j/(n/2)} f_{2j+1}.$$

The language of matrix factorizations can be used to give a higher level description. If $n = 2m$, the matrix F_n can be factorized as

$$F_n \Pi_n = \begin{bmatrix} I_m & \Omega_m \\ I_m & -\Omega_m \end{bmatrix} \begin{bmatrix} F_m & 0 \\ 0 & F_m \end{bmatrix},$$

where Π_n is a permutation matrix and $\Omega_m = \operatorname{diag}(1, \omega, \ldots, \omega^{m-1})$. This factorization shows that an n-point DFT can be computed from two $n/2$-point transforms, and this reduction is the gist of the radix-2 FFT. The book *Computational Frameworks for the Fast Fourier Transform* by Van Loan [284], from which this factorization is taken, shows how, by using matrix notation, the many variants of the FFT can be unified and made easier to understand.

An extended example of how notation can be improved is given by Gillman in the appendix titled "The Use of Symbols: A Case Study" of *Writing Mathematics Well* [104]. Gillman takes the proof of a theorem by Sierpinski (1933) and shows how simplifying the notation leads to a better proof. Knuth set his students the task of simplifying Gillman's version even further, and four solutions are given in [164, §21].

Mathematicians are always searching for better notation. Knuth [163] describes two notations that he and his students have been using for many years and that he thinks deserve widespread adoption. One is notation for the Stirling numbers. The other is the notation $\mathcal{S}$, where $\mathcal{S}$ is any true-or-false statement. The definition is

$$[\mathcal{S}] = \begin{cases} 1, & \text{if } \mathcal{S} \text{ is true,} \\ 0, & \text{if } \mathcal{S} \text{ is false.} \end{cases}$$

For example, the Kronecker delta can be expressed as $\delta_{ij} = [i = j]$. The square bracket notation will seem natural to those who program; indeed, Knuth adapted it from a similar notation in the 1962 book by Iverson that led to the programming language APL [144, p. 11]. The square bracket notation is used in the textbook *Concrete Mathematics* [116]; that book and Knuth's paper give a convincing demonstration of the usefulness of the notation.

Halmos has these words to say about two of his contributions to mathematical notation [127]:

> My most nearly immortal contributions to mathematics are an abbreviation and a typographical symbol. I invented "iff," for "if and only if"—but I could never believe that I was really its first inventor .... The symbol is definitely not my invention—it appeared in popular magazines (not mathematical ones) before I adopted it, but, once again, I seem to have introduced it into mathematics. It is the symbol that sometimes looks like □, and is used to indicate an end, usually the end of a proof. It is most frequently called the "tombstone," but at least one generous author referred to it as the "halmos".

Table 3.1 shows the date of first use in print of some standard symbols; some of them are not as old as you might expect. Not all these notations met with approval when they were introduced. In 1842 Augustus de Morgan complained (quoted by Cajori [49, p. 328 (Vol. II)]):

> Among the worst of barbarisms is that of introducing symbols which are quite new in mathematical, but perfectly understood in common, language. Writers have borrowed from the Germans the abbreviation $n!$ to signify $1.2.3. \ldots (n-1)n$, which gives their pages the appearance of expressing surprise and admiration that $2, 3, 4$, etc., should be found in mathematical results.

3.6. Words versus Symbols

Mathematicians are supposed to like numbers and symbols, but I think many of us prefer words. If we had to choose between reading a paper dominated by symbols and one dominated by words then, all other things being equal, most of us would choose the wordy paper, because we would expect it to be easier to understand. One of the decisions constantly facing the mathematical writer is how to express ideas: in symbols, in words, or both. I suggest some guidelines.

- Use symbols if the idea would be too cumbersome to express in words, or if it is important to make a precise mathematical statement.

- Use words as long as they do not take up much more space than the corresponding symbols.

Table 3.1. First use in print of some symbols. Sources: [49], [116], [163].

Symbol	Name	Year of publication
∞	infinity	1655 (Wallis)
π	pi (3.14159...)	1706 (Jones)
e	e (2.71828...)	1736 (Euler)
i	imaginary unit ($\sqrt{-1}$)	1794 (Euler)
$\equiv$	congruence	1801 (Gauss)
$n!$	factorial	1808 (Kramp)
$\sum$	summation	1820 (Fourier)
$\binom{n}{k}$	binomial coefficient	1826 (von Ettinghausen)
$\prod$	product	1829 (Jacobi)
∇	nabla	1853 (Hamilton)
δ_{ij}	Kronecker delta	1868 (Kronecker)
$|z|$	absolute value	1876 (Weierstrass)
$O(f(n))$	big oh	1894 (Bachmann)
$\lfloor x \rfloor, \lceil x \rceil$	floor, ceiling	1962 (Iverson)

- Explain in words what the symbols mean if you think the reader might have difficulty grasping the meaning or essential feature.

Here are some examples.

(1) Define $C \in \mathbb{R}^{n \times n}$ by the property that vec(C) is the eigenvector corresponding to the smallest eigenvalue in magnitude of A, where the vec operator stacks the columns of a matrix into one long vector.

To make this definition using equations takes much more space, and is not worthwhile unless the notation that needs to be introduced (in this case, a name for $\min\{ |\lambda| : \lambda \text{ is an eigenvalue of } A \}$) is used elsewhere. A possible objection to the above wordy definition of vec is that it does not specify in which order the columns are stacked, but that can be overcome by appending "taking the columns in order from first to last".

(2) Since $|g'(0)| > 1$, zero is a repelling fixed point, so x_k does not tend to zero as $k \to \infty$.

An alternative is

> Since $|g'(0)| > 1$, 0 is a repelling fixed point, so $x_k \nrightarrow 0$ as $k \to \infty$.

This sentence is only slightly shorter than the original and is harder to read—the symbols are beginning to intrude on the grammatical structure of the sentence.

> (3) If $B \in \mathbb{R}^{n \times n}$ has a unique eigenvalue λ of largest modulus then $B^k \approx \lambda^k x y^T$, where $Bx = \lambda x$ and $y^T B = \lambda y^T$ with $y^T x = 1$.

The alternative of "where x and y are a right and left eigenvector corresponding to λ, respectively, and $y^T x = 1$" is cumbersome.

> (4) Under these conditions the perturbed least squares solution $x + \Delta x$ can be shown to satisfy

$$\frac{\|\Delta x\|_2}{\|x\|_2} \leq \epsilon \kappa_2(A) \left(1 + \frac{\|b\|_2}{\|A\|_2 \|x\|_2} \right) + \epsilon \, \kappa_2(A)^2 \frac{\|r\|_2}{\|A\|_2 \|x\|_2} + O(\epsilon^2).$$

> Thus the sensitivity of x is measured by $\kappa_2(A)$ if the residual r is zero or small, and otherwise by $\kappa_2(A)^2$.

Here, we have a complicated bound that demands an explanation in words, lest the reader overlook the significant role played by the residual r.

> (5) If $y_1, y_2, \ldots, y_n$ are all $\neq 1$ then $g(y_1, y_2, \ldots, y_n) > 0$.

In the first sentence "all $\neq 1$" is a clumsy juxtaposition of word and equation and most writers would express the statement differently. Possibilities include

> If $y_i \neq 1$ for $i = 1, 2, \ldots, n$, then $g(y_1, y_2, \ldots, y_n) > 0$.
> If none of the y_i ($i = 1, 2, \ldots, n$) equals 1, then $g(y_1, y_2, \ldots, y_n) > 0$.

If the condition were "$\neq 0$" instead of "$\neq 1$", then it could simply be replaced by the word "nonzero". In cases such as this, the choice between words and symbols in the text (as opposed to in displayed equations) is a matter of taste; good taste is acquired by reading a lot of well-written mathematics.

The symbols $\forall$ and $\exists$ are widely used in handwritten notes and are an intrinsic part of the language in logic. But generally, in equations that are in-line, they are better replaced by the equivalent words "for all" and "there exists". In displayed equations either the symbol or the phrase is acceptable, though I usually prefer the phrase. Compare

$$\sigma(G(t)) = \exp(t\sigma(A)) \quad \text{for all } t \geq 0$$

with

$$\sigma(G(t)) = \exp(t\sigma(A)) \quad \forall t \geq 0.$$

Similar comments apply to the symbols $\Rightarrow$ (implies) and $\Longleftrightarrow$ (if and only if), though these symbols are more common in displayed formulas.

Of course, for some standard phrases that appear in displayed formulas, there is no equivalent symbol:

$$\text{minimize} \quad c^T x - \mu \sum_{i=1}^{n} \ln x_i \quad \text{subject to } Ax = b,$$

$$\underline{\lim} \tfrac{1}{n} \log D_j^n < 0 \quad \text{almost surely,}$$

$$z^T y = \|z\|_D \|y\| = 1, \quad \text{where} \quad \|z\|_D = \max_{v \neq 0} \frac{|z^T v|}{\|v\|}.$$

3.7. Displaying Equations

An equation is displayed when it needs to be numbered, when it would be hard to read if placed in-line, or when it merits special attention, perhaps because it contains the first occurrence of an important variable. The following extract gives an illustration of what and what not to display.

Because $\delta(\bar{x}, \mu)$ is the smallest value of $\|\overline{X}z/\mu - e\|$ for all vectors y and z satisfying $A^T y + z = c$, we have

$$\delta(\bar{x}, \mu) \leq \|\tfrac{1}{\mu}\overline{X}z - e\|.$$

Using the relations $z = \mu X^{-1}s$ and $\bar{x}_i = 2x_i - x_i s_i$ gives

$$\frac{1}{\mu}\overline{X}z = \overline{X}X^{-1}s = (2X - XS)X^{-1}s = 2s - S^2 e.$$

Therefore, $\delta(\bar{x}, \mu) \leq \|2s - S^2 e - e\|$, which means that

$$\delta(\bar{x}, \mu)^2 \leq \sum_{i=1}^{n} (2s_i - s_i^2 - 1)^2 = \sum_{i=1}^{n} (s_i - 1)^4 \leq \left(\sum_{i=1}^{n} (s_i - 1)^2\right)^2 = \delta(x, \mu)^4.$$

The condition $\delta(x, \mu) < 1$ thus ensures that the Newton iterates $\overline{x}$ converge quadratically.

The second and third displayed equations are too complicated to put in-line. The first $\delta(\overline{x}, \mu)$ inequality is displayed because it is used in conjunction with the second display and it is helpful to the reader to display both these steps of the argument. The consequent inequality $\delta(\overline{x}, \mu) \leq \|2s - S^2 e - e\|$ fits nicely in-line, and since it is used immediately it is not necessary to display it.

When a displayed formula is too long to fit on one line it should be broken before a binary operation. Example:

$$|e_{m+1}| \leq |G^{m+1} e_0| + c_n u(1 + \theta_x)\{c(A)|(I - G)^D M^{-1}|$$
$$+ (m + 1)|(I - E) M^{-1}|\}(|M| + |N|)|x|.$$

The indentation on the second line should take the continuation expression past the beginning of the left operand of the binary operation at which the break occurred, though, as this example illustrates, this is not always possible for long expressions. A formula in the text should be broken after a relation symbol or binary operation symbol, not before.

3.8. Parallelism

Parallelism should be used, where appropriate, to aid readability and understanding. Consider this extract:

The Cayley transform is defined by $C = (A - \theta_1 I)^{-1}(A - \theta_2 I)$. If λ is an eigenvalue of A then

$$(\lambda - \theta_2)(\lambda - \theta_1)^{-1}$$

is an eigenvalue of C.

The factors in the eigenvalue expression are presented in the reverse order to the factors in the expression for C. This may confuse the reader, who might, at first, think there is an error. The two expressions should be ordered in the same way.

Parallelism works at many levels, from equations and sentences to theorem statements and section headings. It should be borne in mind throughout the writing process. If one theorem is very similar to another, the statements should reflect that—the wording should not be changed just for the sake of variety (see *elegant variation*, §4.15). However, it is perfectly acceptable to economize on words by saying, in Theorem 2 (say) "Under the conditions of Theorem 1".

For a more subtle example, consider the sentence

It is easy to see that $f(x,y) > 0$ for $x > y$.

In words, this sentence is read as "It is easy to see that $f(x,y)$ is greater than zero for x greater than y." The first $>$ translates to "is greater than" and the second to "greater than", so there is a lack of parallelism, which the reader may find disturbing. A simple cure is to rewrite the sentence:

It is easy to see that $f(x,y) > 0$ when $x > y$.
It is easy to see that if $x > y$ then $f(x,y) > 0$.

3.9. Dos and Don'ts of Mathematical Writing

Punctuating Expressions

Mathematical expressions are part of the sentence and so should be punctuated. In the following display, all the punctuation marks are necessary. (The second displayed equation might be better moved in-line.)

The three most commonly used matrix norms in numerical analysis are particular cases of the Hölder p-norm

$$\|A\|_p = \max_{x \neq 0} \frac{\|Ax\|_p}{\|x\|_p}, \qquad A \in \mathbb{R}^{m \times n},$$

where $p \geq 1$ and

$$\|x\|_p = \left(\sum_{i=1}^{n} |x_i|^p \right)^{1/p}.$$

Otiose Symbols

Do not use mathematical symbols unless they serve a purpose. In the sentence "A symmetric positive definite matrix A has real eigenvalues" there is no need to name the matrix unless the name is used in a following sentence. Similarly, in the sentence "This algorithm has $t = \log_2 n$ stages", the "$t =$" can be omitted unless t is defined in this sentence and used immediately. Watch out for unnecessary parentheses, as in the phrase "the matrix $(A - \lambda I)$ is singular."

Placement of Symbols

Avoid starting a sentence with a mathematical expression, particularly if a previous sentence ended with one, otherwise the reader may have difficulty parsing the sentence. For example, "A is an ill-conditioned matrix" (possible confusion with the word "A") can be changed to "The matrix A is ill-conditioned."

Separate mathematical symbols by punctuation marks or words, if possible, for the same reason.

Bad: If $x > 1$ $f(x) < 0$.

Fair: If $x > 1$, $f(x) < 0$.

Good: If $x > 1$ then $f(x) < 0$.

Bad: Since $p^{-1} + q^{-1} = 1$, $\|\cdot\|_p$ and $\|\cdot\|_q$ are dual norms.

Good: Since $p^{-1} + q^{-1} = 1$, the norms $\|\cdot\|_p$ and $\|\cdot\|_q$ are dual.

Bad: It suffices to show that $\|H\|_p = n^{1/p}$, $1 \le p \le 2$.

Good: It suffices to show that $\|H\|_p = n^{1/p}$ for $1 \le p \le 2$.

Good: It suffices to show that $\|H\|_p = n^{1/p}$ $(1 \le p \le 2)$.

Bad: For $n = r$ (2.2) holds with $\delta_r = 0$.

Good: For $n = r$, (2.2) holds with $\delta_r = 0$.

Good: For $n = r$, inequality (2.2) holds with $\delta_r = 0$.

"The" or "A"

In mathematical writing the use of the article "the" can be inappropriate when the object to which it refers is (potentially) not unique or does not exist. Rewording, or changing the article to "a", usually solves the problem.

Bad: Let the Schur decomposition of A be QTQ^*.

Good: Let a Schur decomposition of A be QTQ^*.

Good: Let A have the Schur decomposition QTQ^*.

Bad: Under what conditions does the iteration converge to the solution of $f(x) = 0$?

Good: Under what conditions does the iteration converge to a solution of $f(x) = 0$?

Notational Synonyms

Sometimes you have a choice of notational synonyms, one of which is preferable. In the following examples, the first of each pair is, to me, the more aesthetically pleasing or easier to read (a capital letter denotes a matrix).

$$\left(\sum_{i,j}(a_{ij}-b_{ij})^2\right)^{1/2}, \qquad \sqrt{\sum_{i,j}(a_{i,j}-b_{i,j})^2},$$

$$\exp\left(2\pi i(x^2+y^2)^{-1/2}\right), \qquad e^{\frac{2\pi i}{\sqrt{x^2+y^2}}},$$

$$(1-n\epsilon)^{-1}|L||U|, \qquad \frac{|L||U|}{1-n\epsilon},$$

$$X_{k+1}=\tfrac{1}{2}X_k(3I-X_k^2), \qquad X_{k+1}=\frac{X_k}{2}[3I-X_k^2],$$

$$\min\{\,\epsilon:|b-Ay|\le\epsilon|A||y|\,\}, \qquad \min\{\,\epsilon\mid|b-Ay|\le\epsilon|A||y|\,\},$$

$$\min\{\,\|A-UBP\|:U^TU=I,\ P\text{ a permutation}\,\},$$

$$\min_{\substack{U^TU=I\\P\text{ a permutation}}}\ \|A-UBP\|.$$

In the next two examples, the first form is preferable because it saves space without a loss of readability.

$$x=\begin{bmatrix}x_1, & x_2, & \cdots, & x_n\end{bmatrix}^T, \qquad x=\begin{bmatrix}x_1\\x_2\\\vdots\\x_n\end{bmatrix},$$

$$\Lambda=\mathrm{diag}(\lambda_i), \qquad \Lambda=\begin{bmatrix}\lambda_1 & & & \\ & \lambda_2 & & \\ & & \ddots & \\ & & & \lambda_n\end{bmatrix}.$$

Of course, the diag($\cdot$) notation should be defined if it is not regarded as standard.

Referencing Equations

When you reference an earlier equation it helps the reader if you add a word or phrase describing the nature of that equation. The aim is to save the reader the trouble of turning back to look at the earlier equation. For example, "From the definition (6.2) of dual norm" is more helpful than "From (6.2)"; and "Combining the recurrence (3.14) with inequality (2.9)" is more helpful than "Combining (3.14) and (2.9)". Mermin [200] calls this advice the "Good Samaritan Rule". As in these examples, the word added should be something more informative than just "equation" (or the ugly abbreviation "Eq."), and inequalities, implications and lone expressions should not be referred to as equations.

Miscellaneous

When working with complex numbers it is best not to use "i" as a counting index, to avoid confusion with the imaginary unit. More generally, do not use a letter as a dummy variable if it is already being used for another purpose.

Note the difference between the Greek letter epsilon, ϵ, and the "belongs to" symbol $\in$, as in $\|x\| \leq \epsilon$ and $x \in \mathbb{R}^n$. Another version of the Greek epsilon is ε. Note the distinction between the Greek letter π and the product symbol $\prod$.

By convention, standard mathematical functions such as sin, cos, arctan, max, gcd, trace and lim are set in roman type, as are multiple-letter variable names. It is a common mistake to set these in italic type, which is ambiguous. For example, is $tanx$ the product of four scalars or the tangent of x?

In bracketing multilayered expressions you have a choice of brackets for the layers and a choice of sizes, for example $\{\,[\,(\,\{\,[\,($, this ordering being the one recommended by *The Chicago Manual of Style* [58]. Most authors try to avoid mixing different brackets in the same expression, as it leads to a rather muddled appearance.

Write "the kth term", not "the kth term", "the k'th term" or "the k-th term." (It is interesting to note that nth is a genuine word that can be found in most dictionaries.)

A slashed exponent, as in $y^{1/2}$, is generally preferable to a stacked one, as in $y^{\frac{1}{2}}$.

The standard way to express that i is to take the values 1 to n in steps of 1 is to write

$$i = 1, \ldots, n \quad \text{or} \quad i = 1, 2, \ldots, n,$$

where all the commas are required. An alternative notation originating in programming languages such as Fortran 90 and MATLAB is $i = 1{:}n$. For counting down we can write $i = n, n - 1, \ldots, 1$ or $i = n{:}{-}1{:}1$, where the middle integer denotes the increment. This notation is particularly convenient when extended to describe submatrices: $A(i{:}j, p{:}q)$ denotes the submatrix formed from the intersection of rows i to j and columns p to q of the matrix A.

Avoid (or rewrite) tall in-line expressions, such as $\begin{bmatrix} g_1 \\ g_2 \end{bmatrix}$, which can disrupt the line spacing.

There are two different kinds of ellipsis: vertically centred ($\cdots$) and "ground level" or "baseline" ($\ldots$). Generally, the former is used between operators such as $+$, $=$, and $\leq$, and the latter is used between a list of

---

Glossary for Mathematical Writing

1. Without loss of generality = I have done an easy special case.

2. By a straightforward computation = I lost my notes.

3. The details are left to the reader = I can't do it.

4. The following alternative proof of X's result may be of interest = I cannot understand X.

5. It will be observed that = I hope you hadn't noticed that.

6. Correct to within an order of magnitude = wrong.

Adapted from [222].

---

symbols or to indicate a product. Examples:

$$x_1 + x_2 + \cdots + x_n, \quad \sigma_1 \geq \sigma_2 \geq \cdots \geq \sigma_n, \quad \lambda_1, \lambda_2, \ldots, \lambda_n, \quad A_1 A_2 \ldots A_n.$$

An operator or comma should be symmetrically placed around the ellipsis; thus $x_1 + x_2 + \cdots x_n$ and $\lambda_1, \lambda_2, \ldots \lambda_n$ are incorrect.

When an ellipsis falls at the end of a sentence there is the question of how the full stop (or period) is treated. Recommendations vary. *The Chicago Manual of Style* suggests typing the full stop before the three ellipsis points (so that there is no space between the first of the four dots and the preceding character). When the ellipsis is part of a mathematical formula it seems natural to put it before the full stop, but the two possibilities may be visually indistinguishable, as in the sentence

> The Mandelbrot set is defined in terms of the iteration $z_{k+1} = z_k^2 + c$, $k = 0, 1, 2, \ldots$.

A vertically centred dot is useful for denoting multiplication in expressions where terms need to be separated for clarity:

$$16046641 = 13 \cdot 37 \cdot 73 \cdot 457,$$

$$\mathrm{cond}(A, x) = \frac{\| \, |I - A^+ A| \cdot |A^T| \cdot |A^{+^T} x| \, \|}{\|x\|}.$$

Care is needed to avoid ambiguity in slashed fractions. For example, the expression $-(b-a)^3/12 f''(\eta)$ is better written as $-\big((b-a)^3/12\big) f''(\eta)$ or $-f''(\eta)(b-a)^3/12$.

Chapter 4
English Usage

There's almost no more beautiful sight than a simple declarative sentence.
— WILLIAM ZINSSER, *Writing with a Word Processor* (1983)

Quite aside from format and style,
mathematical writing is supposed to say something.
Put another way: the number of ideas divided by the
number of pages is supposed to be positive.
— J. L. KELLEY, *Writing Mathematics* (1991)

Let us not deceive ourselves.
"There is no God but Allah" is a more gripping sentence than
Mohammed (*also Mahomet, Muhammad; 570?–632*) *asserted a doctrine of*
unqualified monotheism (*suras 8, 22, 33–37, 89, 91, Koran*).
— MARY-CLAIRE VAN LEUNEN, *A Handbook for Scholars* (1992)

I am about to—or I am going to—die;
either expression is used.
— DOMINIQUE BONHOURS[4] (on his deathbed)

One should not aim at being possible to understand,
but at being impossible to misunderstand.
— QUINTILLIAN

---

[4]Quoted in [42, p. 146].

In this chapter I discuss aspects of English usage that are particularly relevant to mathematical writing. You should keep three things in mind as
you read this chapter. First, on many matters of English usage rules have
exceptions, and, moreover, not all authorities agree on the rules. I have
consulted several usage guides (those described in §2.2) and have tried to
give a view that reflects usage in writing today. Second, about half the
topics discussed here are not peculiar to the English language, but simply
reflect common sense in writing. Third, many of the points mentioned are
not vitally important when taken in isolation. But, as van Leunen explains
(quoted in [164, p. 97]),

> Tone is important, and tone consists entirely of making these
> tiny, tiny choices. If you make enough of them wrong ... then
> you won't get your maximum readership. The reader who has
> to read the stuff will go on reading it, but with less attention,
> less commitment than you want. And the reader who doesn't
> have to read will stop.

4.1. A or An?

Whether *a* or *an* should precede a noun depends on how the first syllable
is pronounced: *a* is used if the first syllable begins with a consonant sound
and *an* if it begins with a vowel sound. For this rule, the initial "yew"
sound in the words university and European is regarded as a consonant
sound: thus "a university", "a European". For words beginning with h, *a*
is used unless the h is not sounded. The only words in this last category
are *heir, honest, honour* (US *honor*), *hour* and their derivatives.

The question "*a* or *an*?" most frequently arises with acronyms, abbreviations and proper nouns. An easy example is "an NP-complete problem".
In the world of mathematical software, given that the usual pronunciation
of LAPACK is l-a-pack, and that of NAG is nag, we would write "an LA-
PACK routine" but "a NAG library routine".

4.2. Abbreviations

One school of thought says that the use of the abbreviations *e.g.* and *i.e.* is
bad style, and that *for example* and *that is* make for a smoother flowing
sentence. In any case, *i.e.* and *that is* should usually be followed by a
comma, and all four forms should be preceded by a comma. When using
the abbreviations you should type *e.g.* and *i.e.*, not *eg.* or *ie.*, since the abbreviations represent two words (the Latin *exempli gratia* and *id est*). Note

that in the following sentence *i.e.* should be deleted: "The most expensive method, i.e. Newton's method, converges quadratically."

The less frequently used *cf.* has only one full stop (a "period" in American English), because it is an abbreviation of a single word: the Latin *confer*, meaning compare. Often, cf. is used incorrectly in the sense of "see", as in "cf. [6] for a discussion". The abbreviation *et al.* is short for *et alia*, so it needs only one full stop.

The abbreviation *N.B.* (of the Latin *nota bene*) is not often used in technical writing, probably because it has to appear at the beginning of a sentence and is somewhat inelegant. You can usually find a better way to give the desired emphasis.

The abbreviation *iff*, although handy in notes, is usually spelled out as *if and only if.*

The normal practice when introducing a nonstandard abbreviation or acronym is to spell out the word or phrase in full on its first occurrence and place the abbreviation immediately after it in parentheses. Thereafter the abbreviation is used. Example:

> Gaussian elimination (GE) is a method for solving a system of n linear equations in n unknowns. GE has a long history; a variant of it for solving systems of three equations in three unknowns appears in the classic Chinese work "Chiu-Chang Suan-Shu", written around 250 B.C.

4.3. Absolute Words

Certain adjectives have an absolute meaning and cannot be qualified by words such as less, quite, rather and very. For example, it is wrong to write *most unique* (replace by *unique*, or perhaps *most unusual*), *absolutely essential*, *more ideal*, or *quite impossible*. However, *essentially unique* is an acceptable term in mathematical writing: it means unique up to some known transformations. Many other words are frequently used with an absolute sense but can be modified (although for some words this usage is open to criticism); example phrases are "convergence is *almost certain*", "a *very complete* survey", "the *most obvious* advantage", and "the function is differentiable *almost everywhere.*"

4.4. Active versus Passive

Prefer the active to the passive voice (prefer "X did Y" to "Y was done by X"). The active voice adds life and movement to writing, whereas too much of the passive voice weakens the communication between writer and reader.

> Passive: The answer was provided to sixteen decimal places by Gaussian elimination.
>
> Active: Gaussian elimination gave the answer to sixteen decimal places.

> Passive: The failure of Newton's method to converge is attributed to the fact that the Jacobian is singular at the solution.
>
> Active: Newton's method fails to converge because the Jacobian is singular at the solution.

> Passive: A numerical example is now given to illustrate the above result.
>
> Active: We give a numerical example to illustrate this result, *or* The following numerical example illustrates this result.

The second example in the following trio illustrates a further degree of abstraction in which a verb is replaced by an abstract noun modified by another verb.

> Passive: The optimality of y was verified by checking that the Hessian matrix was positive definite.
>
> Passive and indirect: Verification of the optimality of y was achieved by checking that the Hessian matrix was positive definite.
>
> Active: We verified the optimality of y by checking that the Hessian matrix was positive definite.

Other "was" phrases that can often be removed by rewriting are "was performed", "was experienced", "was carried out", "was conducted" and "was accomplished".

The passive voice has an important role to play, however. It adds variety, may be needed to put emphasis on a certain part of a sentence, and may be the only choice if specific information required for an active variant is unknown or inappropriate to mention. Examples where the passive voice allows the desired emphasis are "An ingenious proof of this conjecture was constructed by C. L. Ever", and (from the writings of Halmos [245, p. 96]) "The subjects just given honorable mention, as well as the three actually discussed in detail, have been receiving serious research attention in the course of the last twenty years." The passive voice is also useful for euphemistic effect, allowing the clumsy experimenter to say "The specimen was accidentally strained during mounting" instead of "I dropped the specimen on the floor."

The Ten Commandments of Good Writing

1. Each pronoun should agree with their antecedent.

2. Just between you and I, case is important.

3. A preposition is a poor word to end a sentence with.

4. Verbs has to agree with their subject.

5. Don't use no double negatives.

6. Remember to never split an infinitive.

7. When dangling, don't use participles.

8. Join clauses good, like a conjunction should.

9. Don't write a run-on sentence it is difficult when you got to punctuate it so it makes sense when the reader reads what you wrote.

10. About sentence fragments.

Reprinted, with permission, from *How to Write and Publish a Scientific Paper* [68].

4.5. Adjective and Adverb Abuse

Use an adjective or adverb only if it earns its place. The adjectives or adverbs *very, rather, quite, nice* and *interesting* should be used with caution in technical writing, as they are imprecise. For example, in "The proof is very easy" and "This inequality is quite important" the adverbs are best omitted. Examples of acceptable usage are "These results are very similar to those of Smith" and "This bound can be very weak." *Interesting* is an overworked adjective that can often be avoided. For example, in the sentence "It is interesting to re-prove this result using Laplace transforms", *instructive* is probably the intended word.

Try to avoid using nouns as adjectives. "The method for iteration parameter estimation" can be expressed more clearly as "The method for estimating iteration parameters." While proper nouns are often used as adjectives in speech ("this sequence is Cauchy", "that matrix is Toeplitz"), such usage in formal writing is best avoided (write "this is a Cauchy sequence", "that is a Toeplitz matrix"). Similarly, write "Euler's method is unstable" instead of "Euler is unstable."

An *adverb* that is overworked in mathematical writing is *essentially*. Dictionaries define it to mean necessarily or fundamentally, but it is often used with a vague sense meaning "almost, but not quite". Before using the word, consider whether you can be more precise.

Bad: Beltrami (1873) essentially derived the singular value decomposition.

Good: Beltrami (1873) derived the singular value decomposition for square, nonsingular matrices.

A valid use of *essentially* is in the expression "essentially the same as", which by convention in scientific writing means "the same, except for minor details".

4.6. -al and -age

Certain words that can be extended with an -al or -age ending are often misused in the extended form. The suffix tends to give a more abstract meaning, which makes it more difficult to use the word correctly. For example, *usage* means a manner of using, so correct *usage* is illustrated by "in the original usage the conjugate gradient method was not preconditioned" and "the use of Euler's method is not recommended for stiff differential equations."

An example where an -al ending is used incorrectly is "the most pragmatical opinion is the one expressed by the term's inventor", in which the third word should be "pragmatic".

4.7. Ambiguous "This" and "It"

A requirement of good writing is to make clear to the reader, at all times, what is the entity under discussion. *This* phrases such as "This is a consequence of Theorem 2" should be used with caution as they can force the reader to backtrack to find what "this" refers to. Often it helps to insert an appropriate noun after *this*. *It* can also be ambiguous: in the sentence "Condition 3 is not satisfied for the steepest descent method, which is why we do not consider it further" we cannot tell whether it is the condition or the method that is not being pursued.

4.8. British versus American Spelling

In my opinion (as a Briton) it makes little difference whether you use British or American (US) spellings, as long as you are consistent within a given piece of writing. For some journals, copy editors will convert a manuscript to one or the other form of spelling. Major dictionaries give both spellings. I find it natural to use the spelling of the country in which I am working at the time! See §5.9 for some examples of the different spellings.

4.9. Capitalization

Words that are derived from a person's name inherit the capitalization. Thus: Gaussian elimination, Hamiltonian system, Hermitian matrix, Jacobian matrix, Lagrangian function, Euler's method, and so on. The incorrect form "hermitian" is sometimes seen. The Lax Equivalence Theorem is quite different from a lax equivalence theorem! Some (but not all) dictionaries list "abelian" with a small "a", showing that eponymous adjectives can gradually become accepted in uncapitalized form.

There does not seem to be a standard rule for when to capitalize the word following a colon. Bernstein [28] and Knuth [164, p. 11] both suggest the useful convention of capitalizing when the phrase following the colon is a full sentence.

4.10. Common Misspellings or Confusions

The errors shown in Table 4.1 seem to be common in mathematical writing. The misspellings marked with an asterisk are genuine words, but have different meanings from the corresponding words in the left column.

One Web page I visited describes "seperate" as the most common misspelling on the Internet and lose/loose as the second most common. Using the Web search engine Alta Vista I found one occurrence of "seperate" for every 24 occurrences of "separate".

The journal *Physical Review Letters* started spelling *Lagrangian* as *Lagrangean* in mid 1985, a change which is incorrect according to most dictionaries. Mermin, a Cornell physicist, spotted the switch and wrote an article criticizing it [201]. The journal has now reverted to *Lagrangian*.

According to McIlroy [199], on most days at Bell Laboratories someone misspells the word *accommodate*, in one of seven incorrect ways.

4.11. Consistency

It is important to be consistent. Errors of consistency often go unnoticed, but can be puzzling to the reader. Don't refer to ker(A) as the null-space, or null(A) as the kernel—stick to matching synonyms. If you use the term "Cholesky factorization", don't say the "Cholesky decomposition" in the

Table 4.1. Common errors. Asterisk denotes a genuine word.

Correct/Intended	Misspelling
analogous	analagous
criterion	criteria*
dependent	dependant*
discrete	discreet*
Frobenius	Frobenious
idiosyncrasy	idiosyncracy
in practice	in practise
led (past tense of lead)	lead*
lose	loose* (very common)
phenomenon	phenomena* (plural of phenomenon)
preceding	preceeding
principle	principal*
propagation	propogation
referring	refering
Riccati	Ricatti
separate	seperate
supersede[a]	supercede
zeros	zeroes[b]

[a]The only English word ending in -sede.

[b]Some, but not all, dictionaries give the -oes ending as an alternative spelling for the plural noun.

same work. If words have alternative spellings, stick to one: don't use both orthogonalise and orthogonalize, and if you use orthogonalize also use, for example, optimize, not optimise. But note that not all -ise words can be spelt with -ize; examples are listed in §5.9.

4.12. Contractions

Contractions such as *it's*, *let's*, *can't* and *don't* are not used in formal works, but are acceptable in popular articles if used sparingly. Note the distinction between the contraction *it's* (short for *it is*) and the possessive *its*: "It's raining", "A matrix is singular if its determinant is zero." One editor comments that the two most frequent errors she encounters are the use of *it's* for *its* and incorrect punctuation surrounding *however* [286, p. 39].

4.13. Dangling Participle

What is wrong with the following sentence?

Substituting (3) into (7), the integral becomes $\pi^2/9$.

This sentence suggests that the integral makes the substitution. The error is that the intended subject ("we") of the participle *substituting* is not present in the sentence. Rewritten and unambiguous versions are

Substituting (3) into (7), we find that the integral is $\pi^2/9$.
When (3) is substituted into (7), the integral becomes $\pi^2/9$.

A similar example that is less obviously wrong is

When deriving parallel algorithms, the model of computation must be considered carefully.

Dangling participles are usually not ambiguous when read in context, but they can be distracting:

A bug was found in the program using random test data.

Here is another example of a different type:

Being stiff, the Runge-Kutta routine required a large amount of CPU time to solve the differential equation.

Here, the problem is that the noun immediately following "being" is not the one to which this participle refers. There are several ways to rewrite the sentence. One that preserves the emphasis is

Because the differential equation is stiff, the Runge-Kutta routine required a large amount of CPU time to solve it.

Certain participial constructions are idiomatic and hence are regarded as acceptable:

Assuming $G(x^*)$ is positive definite, x^* is a minimum point for F.
Considering the large dimension of the problem, convergence was obtained in remarkably few iterations.
Strictly speaking, the bound holds only for $n\epsilon < 1$.

4.14. Distinctions

Affect, Effect. *Affect* is a verb meaning to produce a change. *Effect* is a noun meaning the result of a change. Examples: "Multiple roots affect the convergence rate of Newton's method", "One effect of multiple roots is to reduce the convergence rate of Newton's method to linear." Effect is also a verb meaning to bring about (as in "to effect a change"), but in this form it is rarely needed in mathematical writing.

Alternate, Alternative. *Alternate* implies changing repeatedly from one thing to another. An *alternative* is one of several options. Compare "While writing his thesis the student alternated between elation and misery", with "The first attempt to prove the theorem failed, so an alternative method of proof was tried."

Compare with, Compare to. *Compare with* analyses similarities and differences between two things, whereas *compare to* states a resemblance between them. Examples: "We now compare Method A with Method B", "Shakespeare compared the world to a stage", "Shall I compare thee to a summer's day?" As Bryson [41] explains, "Unless you are writing poetry or love letters, *compare with* is usually the expression you want." *Compare and* is an alternative to *compare with*: "We now compare Method A and Method B" or, better, "We now compare Methods A and B."

Compose, Comprise, Constitute. *Compose* means to make up, *comprise* means to consist of. "Comprised of" is always incorrect. Thus, "the course is composed of three topics" or "the course comprises three topics", but not "the course is comprised of three topics." *Constitute* is a transitive verb used in the reverse sense: "these three topics constitute the course."

Due to, Owing to. These two expressions are not interchangeable, though writers frequently use *due to* in place of *owing to*. Use *due to* where you could use "caused by", or "attributable to"; use *owing to* where you could use "because of". Thus "The instability is due to a rank deficient submatrix" but "Owing to a rank deficient submatrix the computed result was inaccurate."

Fewer, Less. *Less* refers to quantity, amount or size, *fewer* to number. Thus "the zeros of $f(x)$ are less than those of $g(x)$" means that if x is a zero of f and y a zero of g then $x < y$, whereas "the zeros of $f(x)$ are fewer than those of $g(x)$" means that g has more zeros than f. Bryson [41] states the rule of thumb that *less* should be used with singular nouns and *fewer* with plural nouns: less research, less computation, fewer graduates, fewer papers.

Practice, Practise. In British English, *practice* is the noun and *practise* the verb (as with advice and advise). Thus "in practice", "practice session", "practise the technique", "practised speaker". But in American English both verb and noun are spelt *practice*.

Which, That. A "wicked which"[5] is an instance of the word *which* that should be *that* (example: replace the word before *should*, earlier in this sentence, by *which*). The rule is that *that* defines and restricts, whereas *which* informs and does not restrict.[6] (Mathematicians should be good at spotting definitions.) Note the difference between the following two examples.

"Consider the Pei matrix, which is positive definite." We are being told additional information about the Pei matrix: that it is positive definite.

"Consider the Pei matrix that is positive definite." Now we are being asked to focus on a particular Pei matrix from among several: the one that is positive definite.

A useful guide is that which-clauses are surrounded by commas, or preceded by a comma if at the end of a sentence. If you're not sure whether to use *which* or *that*, see whether your sentence looks right with commas around the relevant clause. Sometimes it pays to introduce a wicked which to avoid ugly repetition, as has been done in the sentence "This approach is similar to that which we used in our earlier paper" (though "the one we used" is better). A rule of thumb discussed in [164, pp. 96–97] is to replace *which* by *that* whenever it sounds right to do so.

4.15. Elegant Variation

Elegant variation is defined by the Fowlers [84] as "substitution of one word for another for the sake of variety". It is a tempting way to avoid repetition, but is often unnecessary and can introduce ambiguity. Consider the sentence "The eigenvalue estimate from Gershgorin's theorem is a crude bound, but it is easy to compute." Does Gershgorin's theorem yield an estimate or a strict bound? We cannot tell from the sentence. In fact, the answer is that it can yield either, depending on how you interpret the theorem. A rewrite of the sentence (with knowledge of the theorem) produces "The eigenvalue inclusion regions provided by Gershgorin's theorem are crude, but easy to evaluate", where *inclusion regions* can be replaced by *estimates* or *bounds*, depending on the desired emphasis.

The opposite of elegant variation is when the same word is repeated in different forms or with different meanings. Here are two examples.

The performance is impressive and gives the impression that

---

[5]A term coined by Knuth [164].

[6]Some authorities permit *which* to be used in a defining clause (e.g., Gowers [115]), but, as Bryson [41] puts it, "the practice is on the whole better avoided."

the blocksize is nearly optimal. [impressive, impression]
In the remainder of this chapter we examine the remainder in
Euler's summation formula. [remainder, remainder]

Such echoing of words is distracting and is easily avoided by choosing a
synonym for one of them.

4.16. Enumeration

Consider the extract

The Basic Linear Algebra Subprograms (BLAS) have several
advantages. They

- Lead to shorter and clearer codes.
- Improve modularity.
- Machine dependent optimizations can be confined to the
  BLAS, aiding portability, and
- Tuned BLAS have been provided by several manufacturers.

This explanation reads badly because the entries in the list are not grammatically parallel: the preceding "they" applies only to the first two entries
of the list, and the third entry is not a complete sentence, unlike the others.
This is an example of bastard enumeration, so-named by Fowler [83, p. 28],
who explains that in an enumeration "there must be nothing common to
two or more of the items without being common to all."

4.17. False If

The if–then construct is a vital tool in expressing technical arguments, but
it is sometimes used incorrectly. Consider the sentence

If we wish to compare the solutions of $f - \lambda k(f) = 0$ and $f_n -
\lambda k_n(f_n) = 0$, then Jones shows that for a wide class of nonlinear
$k(f)$, $\|f - f_n\| \le c(\lambda)\|k_n(f) - k(f)\|$.

Jones's demonstration is independent of whether or not we wish to compare
solutions, so the *if* is misleading: it falsely heralds a logical condition. False
ifs can always be removed by rewriting:

To compare the solutions of $f - \lambda k(f) = 0$ and $f_n - \lambda k_n(f_n) = 0$,
we can use Jones's result that for a wide class of nonlinear $k(f)$,
$\|f - f_n\| \le c(\lambda)\|k_n(f) - k(f)\|$.

A more confusing example is

> If we assume that rational fractions behave like almost all real
> numbers, a theorem of Khintchine states that the sum of the
> first k partial quotients will be approximately $k \log_2 k$.

The *if* appears to be a false one, because the statement of Khintchine's
theorem must be independent of what the writer assumes. In fact, with
knowledge of the theorem, it can be seen that the main error is in the word
"states". If we change "states" to "implies" and delete "we assume that",
then the sentence is correct.

The next example is an unnecessary *if*, rather than a false *if*.

> We show that if $\hat{x}$ is the computed solution to $Lx = b$ then
> $(L + \Delta L)\hat{x} = b$, where $\|\Delta L\| \le \alpha(n)\epsilon\|L\|$.

This type of construction is acceptable if used sparingly. I prefer

> We show that the computed solution $\hat{x}$ to $Lx = b$ satisfies $(L +
> \Delta L)\hat{x} = b$, where $\|\Delta L\| \le \alpha(n)\epsilon\|L\|$.

4.18. Hyphenation

As Turabian [278, p. 44] notes, the trend is not to hyphenate compound
words beginning with prefixes such as multi, pre, post, non, pseudo and
semi. In mathematical writing it is common to write *nonsingular, semidef-
inite* (but *semi-infinite* to avoid a double i) and *pseudorandom*. However, a
hyphen is retained before a proper noun, as in *non-Euclidean*. In deciding
whether to hyphenate or to combine two words as one, it is worth bearing
in mind that the hyphenated form tends to be easier to read because the
prefix can be seen at a glance. Readers whose first language is not English
may appreciate the hyphenated form.

Compound words involving *ill* and *well* occur frequently in mathematical writing and opinions differ about their hyphenation. *The Chicago Manual of Style* [58] recommends hyphenation when a compound with *ill, well,
better, best, little, lesser*, etc., appears before a noun, unless the compound is
itself modified. The purpose of this hyphenation rule is to avoid ambiguity.
Examples:

> This is an ill-posed problem *but* This problem is ill posed.
> The well-known theorem *but* The theorem is well known.
> An ill-conditioned function *but* A very ill conditioned function.
> The second-order term has a constant 2 *but* This term is of
> second order.

The second example is widely accepted, but many writers always hyphenate compounds involving *ill*, such as ill-conditioned, and it is hard to argue against this practice. There are some common phrases that some writers hyphenate and others do not. An example is *floating point arithmetic* (*floating-point arithmetic*).

In the phrase "we use the 1, 2 and ∞-norms", a *suspensive hyphen* is required after "1" and "2" since they are prefixes to "norm" and we need to show that they are to be linked to this word. Thus the phrase should be rewritten "we use the 1-, 2- and ∞-norms".

Notice the hyphen in the title of Halmos's best-seller *Finite-Dimensional Vector Spaces* [122]. Halmos explains in [128, p. 146] that in the original 1942 edition the hyphen was omitted, but it was added for the 1958 edition.

4.19. Linking Words

If a piece of writing is to read well there must be no abrupt changes in mood or direction from sentence to sentence within a paragraph (unless such changes are used deliberately for effect). One way to achieve a smooth flow is to use linking words. Notice how the following paragraph would be improved by adding "In particular" to the start of the second sentence and "Furthermore" to the start of the third.

> Once we move from a convex program to a general nonlinear program, matters become far more complicated. Certain topological assumptions are required to avoid pathological cases. The results apply only in a neighbourhood of a constrained minimizer, and involve convergence of subsequences of global minimizers of the barrier function.

Of course, a sequence of sentences of the form "adverb, fact" quickly becomes tiresome, so linking words should not be overused.

Here is a list of linking words and phrases, arranged according to sense. For examples of use see §5.8.

combinations. also, and, as well as, besides, both, furthermore, in addition to, likewise, moreover, similarly.

implications or explanations. as, because, conversely, due to, for example, given, in other words, in particular, in view of, it follows that, otherwise, owing to, since, specifically, that is, thus, unlike.

modifications and restrictions. although, alternatively, but, despite, except, however, in contrast, in spite of, nevertheless, of course, on the contrary, on the other hand, though, unfortunately, whereas, yet.

emphasis. actually, certainly, clearly, in fact, indeed, obviously, surely.

consequences. accordingly, consequently, hence, therefore, thus.

4.20. Misused and Ambiguous Words

Both. A common misuse of *both* is illustrated by "In Gaussian elimination we can order the inner loops 'ij' or 'ji'. Both orderings are equivalent, mathematically." *Both* means "the two together" and is redundant when the sentence already carries this implication, as in this example. It would also be incorrect to say "Both orderings produce identical results." Correct versions are "These orderings are equivalent, mathematically", "Both orderings yield the same result", or "The two orderings produce identical results." Another common misuse of *both* is misplacement when it is used with prepositional phrases. For example,

> Incorrect: "Solutions are found both in the left and right quadrants." (*Both* is followed by a preposition, *in the left*, but *and* is followed by a noun.)
>
> Correct: "'Solutions are found both in the left and in the right quadrants." (Prepositional phrases follow *both* and *and*.)
>
> Correct: "'Solutions are found in both the left and the right quadrants." (Nouns follow *both* and *and*.)

Like. Consider the sentence

> Solving triangular systems is such a common operation that it has been standardized as a subroutine, along with many other common linear algebra operations like matrix multiplication.

The word *like* incorrectly limits the choice of linear algebra operations rather than serving as an example. Replacing *like* by *such as* conveys the intended meaning. The correct use of *like* is illustrated by "The Schulz iteration is quadratically convergent, like the Newton iteration."

Problem. An overused and, at times, ambiguous word in mathematical writing is *problem*, which can refer to both the focus of a piece of work and the difficulties encountered in carrying out the work. Sentences such as "In solving this problem we encountered a number of problems" can be avoided by using a synonym for the second occurrence of "problem", or rewriting. The sentence "We describe the special problems arising when solving stiff differential equations" is ambiguous: "problems" could refer to classes of sub-problems produced by the solution process (such as nonlinear equations), or particular difficulties faced when solving the differential equations. Again, a rewrite is necessary.

Reason. In the phrase "the reason ... is because" the word *because* is redundant, since it means "for the reason that". Therefore in the sentence "The reason for the slow convergence is because α is a double root" *because* should be replaced by *that*. Similarly, in the phrase "the reason why" the word *why* can often be omitted. Thus "The reason why this question is important is that" is better written as "The reason this question is important is that" or "This question is important because".

Significant. Be careful if you use the word *significant* in mathematical writing, because to some readers it is synonymous with *statistically significant*, which carries a precise statistical meaning.

Try and, try to. The expression *try and* is frequently used in spoken English, but it is colloquial and should be replaced by *try to* in written English. Thus in the sentence "We sum the numbers in ascending order to try and minimize the effect of rounding errors" *try and* should be replaced by *try to*.

4.21. Numbers

Small integers should be spelled out when used as adjectives ("The three lemmas"), but not when used as names or numbers ("The median age is 43" or "This follows from Theorem 3"). The number 1 is a special case, for often "one" or "unity" reads equally well or better: "z has modulus one".

4.22. Omit These Words?

Here are some words and phrases whose omission often improves a sentence:

actually, very, really, currently, in fact, thing, without doubt.

The phrase "we have" can often be omitted. "Hence we have $x = 0$" should be replaced by "Hence $x = 0$." "Hence we have the following theorem" should be deleted or replaced by a sentence conveying some information (e.g., "Hence we have proved the following theorem").

4.23. Paragraphs

A standard device for making text more appetizing is to break it into small paragraphs, as is done in newspapers. The short paragraph principle is worth following in technical writing, where the complexity of the ideas makes it more important than usual for the reader to be given frequent rests. Furthermore, long paragraphs tend to give a page a heavy image that can be a visual deterrent to the potential reader. A mix of different lengths is best. Ideally, each paragraph contains a main idea or thought that separates it from its neighbours. A long paragraph that is hard to break may be indicative of convoluted thinking.

The best writers occasionally slip in one-sentence paragraphs.

4.24. Punctuation

Much can be said about punctuation, and for thorough treatments of the topic I refer the reader to the references mentioned in Chapter 2. It is worth keeping in mind Carey's explanation [52] that "the main function of punctuation is *to make perfectly clear the construction* of the written words." A few common mistakes and difficulties deserve mention here.

- In "This result is well known, see [9]" the comma should be a semi-colon, which conveys a slightly longer pause. Even better is to say "This result is well known [9]." A common mistake is the let-comma-then construction: "Let x^* be a local maximum of $F(x)$, then a Taylor series expansion gives . . . ." The comma should be a full stop.

  Similarly, the comma should be a semicolon, or even a full stop, in this sentence: "This bound has the disadvantage that it uses a norm of X, moreover the multiplicative constant can be large when X is not a normal matrix." These errors are called "comma splices" by Gordon [111].

- In this sentence the semicolon should be a comma: "The secant method can also be used; its lack of need for derivatives being an advantage." A rough guide is to use a semicolon only where you could also use a full stop.

- In the sentence "If we use iterative refinement solutions are computed to higher accuracy", a comma is needed after "refinement", otherwise the reader may take "iterative refinement" as modifying "solutions". Another example where a comma is needed to avoid ambiguity is the sentence "However, the singularity can be removed by a change of variable."

- In sentences such as

  Fortran 77 contains the floating point data types real, complex and double precision.
  The output can be rotated, stretched, reduced or magnified.

we have a choice of whether or not to place a comma (called a serial comma) before the *and* or *or* that precedes the final element of the list. In some sentences a serial comma is needed to avoid ambiguity, as in the sentence "A dictionary is used to check spelling, shades of meaning, and usage", where the absence of the comma makes "shades" modify usage. Opinion differs on whether a serial comma should always be used. It is a matter of personal preference. The house styles of some publishers require serial commas.

If a list contains commas within the items, ambiguity can be avoided by using a semicolon as the list separator. Example:

> The test collection includes matrices with known inverses or known eigenvalues; ill-conditioned or rank deficient matrices; and symmetric, positive definite, orthogonal, defective, involutary, and totally positive matrices.

- The exclamation mark should be used with extreme caution in technical writing. If you are tempted to exclaim, read "!" as "shriek"; nine times out of ten you will decide a full stop is adequate. An example of correct usage is, from [217, p. 46], "When A is tridiagonal the computation of $A^{-1}u$ costs little more than the computation of Au!" The exclamation mark could be omitted, but then the reader might not realize that this is a surprising fact. Another example is, from [159, p. 42], "The chi-square table tells us, in fact, that V_2 *is much too low*: the observed values are so close to the expected values, we cannot consider the result to be random!"

- In the US, standard practice is to surround quotes by double quotation marks, with single quotation marks being used for a quote within a quote. In the UK, the reverse practice is generally used. The placement of final punctuation marks also differs: in US usage, the final punctuation is placed inside the closing quotation marks (except for "!" and "?" when they are not part of the quotation), while in UK usage it goes outside (except for "!" and "?" when they are part of the quotation). In this book, for quotations that end sentences, the end of sentence period appears outside quotation marks unless the quotation is itself a valid sentence.

- An apostrophe denotes possession for nouns (*the proof's length*) but not for pronouns (*this book is yours*). An apostrophe is used with the plural of letters and of words when the words are used without regard to their meaning: "there are three l's in the word parallel", "his prose contains too many however's." For plurals of numbers the apostrophe can be omitted: "a random matrix of 0s and 1s". For plurals of mathematical symbols or expressions the apostrophe can again be omitted provided there is no ambiguity: "these fs are all continuous", "likewise for the Z_ks", "these δs are all of order 10^{-8}."

4.25. Say It Better, Think It Gooder

The title of this section combines the titles of two papers by George Piranian [227] and Paul Halmos [126] that appeared in *The Mathematical Intelligencer* in 1982. As Gillman explains in [105], "George said good English is important. Paul said, what do you mean, good English is important?—good mathematics is more important. They are both right." While correct English usage is important, it must not be allowed to deflect you from the language-independent tasks of planning and organizing your writing.

4.26. Saying What You Mean

In technical writing you need to take great care to say what you mean. A hastily constructed sentence can have a meaning very different from the one intended. In a book review in *SIAM Review* [vol. 34, 1992, pp. 330–331] the reviewer quotes the statement "According to Theorem 1.1, a single trajectory $X(t,x)$ passes through almost every point in phase space ...." The book's author meant to say that for every point in phase space there is a unique trajectory that goes through it.

4.27. Sentence Opening

Try not to begin a sentence with *there is* or *there are*. These forms of the verb *be* make a weak start to a sentence, because they delay the appearance of the main verb (see the quote by Dixon on page 77). Sometimes these phrases can simply be deleted, as in the sentence "There are several methods that are applicable" ("Several methods are applicable"). Also worth avoiding, if possible, are "It is" openers, such as "It is clear that" and "It is interesting to note that". If you can find alternative wordings, your writing will be more fresh and lively.

4.28. Simplification

Each word or phrase in the left column below can (or, if marked with an asterisk, should) be replaced by the corresponding one in the right column. This is not an exhaustive list (see [69, Appendix 4] or [14] for many

Table 4.2. The origins of some synonyms.

Anglo-Saxon	French	Latin
ask	question	interrogate
rise	mount	ascend
good	marvellous	superior
show	establish	exhibit, demonstrate
need	requirement	necessity
think	ponder	cogitate

more examples) but comprises periphrastic phrases that I have spotted in
mathematical writing.

by means of	by
conduct an investigation	investigate
due to the fact that*	since/because
firstly	first
in a position to	can
in order to	to
in the case that*	if
in the course of	during
in the event that*	if
in the first place*	first
it is apparent that	apparently
it may happen that	there may/might
most unique*	unique
take into consideration	consider
that is to say	in other words/that is
to begin with*	to begin

4.29. Synonym Selection

English is one of the most synonym-rich languages, thanks to the words it
has adopted from other languages, and each member of a set of synonyms
can have a different tone and shade of meaning. A dictionary and a thesaurus are vital aids to choosing the right word (see §2.1). When you have
a choice of words, use a short, concrete one in preference to a long or abstract one. Often, this means using an Anglo-Saxon word instead of one of
French or Latin origin, as Table 4.2 illustrates.

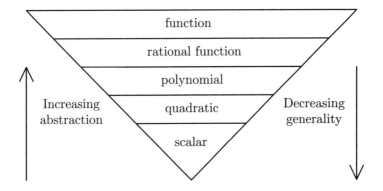

Figure 4.1. Word pyramid.

Don't be afraid of using a long or unusual word if it is just the right one, or if it adds spice to your writing or conveys an interesting image. Acton [3, p. 145] writes "The second term ... is obstreperous." No other word quite so vividly describes the uncontrollable term to which he refers (with the possible exception of *recalcitrant*, another word used by Acton in the same book).

I chose the title "Otiose Symbols" on page 29 in preference to "Unnecessary Symbols" for two reasons. First, the former is shorter and more likely to catch the reader's eye. Second, *otiose* means "serving no practical purpose", so "otiose symbols" has a stronger meaning than "unnecessary symbols".

It helps to think of word pyramids showing levels of abstraction [281], such as the one in Figure 4.1. As Turk and Kirkman [280] point out, "Abstract words are less easy to decode because the reader has to 'scan' all the possibilities subconsciously before deciding on a specific meaning." Generally, it is best to use the least abstract, most specific word possible. This leads to more lively and precise prose. Here are some examples of chains of words in order of increasing specificity:

    result–theorem–inequality
    statistic–error–relative error
    optimum–minimum–global minimum
    random–normally distributed–normal (0,1)

4.30. Tense

One of the decisions the writer of a technical paper must make is what tense to use. There are no hard and fast rules, except to be consistent in the use of tense, but I offer some guidelines.

I prefer the present tense to the future tense for referring to later parts of the paper. Thus I write "This is proved in Theorem 3, below" rather than "This will be proved in Theorem 3", and "We discuss these matters in detail in Section 5" rather than "We will discuss these matters in detail." Note that since the latter example does not contain a section reference it could be construed as referring to a future paper; this potential ambiguity is a danger of using the future tense.

The present tense can also be used for backward references, as in "The analysis of Section 1 proves the following ...", though the past tense is more common: "We showed/saw in Section 1 that ...." Tables and figures exist in the present, so the present tense should be used to refer to them: "Table 4 shows that", "Figure 1 displays". Similarly, in a citation where the reference is the subject of the sentence, the present tense is the one to use: "Reference [4] contains the interesting result that ...."

To refer to work in earlier papers either the past or the present tense can be used: "Banach shows that" or "Banach showed that". I use the present tense unless I want to emphasize the historical aspects of the reference, as, for example, when summarizing a number of earlier results in a survey.

A summary or conclusions section should use the past tense when specifying actions: "We have shown that" or "We showed that" rather than "We show that". But if a simple statement of results is given, the present tense is appropriate: "Asymptotic expansions are useful for ...", "Our conclusion is that ...."

In summary, I recommend the rule "if in doubt use the present tense."

4.31. What to Call Yourself

All technical writers face the question of how to refer to themselves in their papers. For example, if you refer to a previous result of yours you may have to choose between

(1) I showed in [3] that
(2) We showed in [3] that
(3) The author showed in [3] that
(4) It was shown in [3] that

Sentence 4 should be avoided because it is in the passive voice, which is unnecessary in this case. Sentence 3 has a formal, stilted air and is again best avoided. Some authorities on technical writing advise against using "we" to refer to a single author, but are happy for it to be used when there are two or more authors. Nevertheless, in mathematical writing "we" is by far the most common choice of personal pronoun; I frequently use it myself (in papers, but not in this book). "I" creates a more frank, personal contact with the reader and is less commonly used than "we". Zinsser [304, p. 24] suggests "If you aren't allowed to use 'I', at least think 'I' while you write, or write the first draft in the first person and then take the 'I's out. It will warm up your impersonal style."

"We" can be used in the sense of "the reader and I", as in the sentence

We saw earlier that $f_n \to 0$ as $n \to \infty$.

Whether you choose "I" or "we", you should not mix the two in a single document, except, possibly, when using the "reader and I" form of "we".

Other personal pronouns to consider are "one" and "us". "One", as in "one can show that ..." is often used, but is perhaps best avoided because of its vague, impersonal nature. "Us" is useful in sentences such as

Let us now see how the results can be extended to non-smooth functions.

where it means "the reader and I". It is generally quite easy to remove both "one" and "us" by phrasing sentences in an alternative form. For example, the sentence above can be rewritten as "How can the results be extended to non-smooth functions? One approach is ...", or "To see how the results can be extended to non-smooth functions, we ...."

4.32. Word Order

The first words of a sentence are usually regarded by the reader as the most important. Therefore, the word order of sentences should be chosen to reflect the desired emphasis. Compare the first sentence of this paragraph with "The reader usually regards the first words of a sentence as the most important", which places the emphasis on the reader rather than "the first words of a sentence".

Reordering the words of a sentence can strengthen it and remove ambiguity. Here is an example of a misplaced *only*.

> Bad: The limit point is only a stationary point when the regularity conditions are satisfied. (Might be taken to mean that the limit point is expected to be more than just a stationary point.)

Good: The limit point is a stationary point only when the regularity conditions are satisfied.

Here is an example of an incorrect *either*.

Bad: Mathematically, either we have an integral equation of the first kind or one of the second kind.

Good: Mathematically, we have an integral equation of either the first kind or the second kind. (Or simply delete *either* from the bad example.)

Chapter 5
When English Is a Foreign Language

Two authors of the present book ...
freely acknowledge that they are more fluent in German and French,
yet they enjoy communicating in English and are convinced that it is
more than mere historical coincidence ... that has led English
to become the lingua franca *of modern science.*

— HANS F. EBEL, CLAUS BLIEFERT, and WILLIAM E. RUSSEY,
The Art of Scientific Writing (1987)

I ventured to write this book in English because
it will be more easily read in poor English
than in good German by 90% of my intended readers.

— HANS J. STETTER, *Analysis of Discretization Methods for*
Ordinary Differential Equations (1973)

England and America are two countries divided by a common language.

— GEORGE BERNARD SHAW

Please feel free to change any words that you wish.
I tried to make it sound Vulcan—a lot of unnecessarily long words.

— Star Trek: The Next Generation, "Unification II", Stardate 45245.8.

English grammar is so complex and confusing
for the one very simple reason that
its rules and terminology are based on Latin—
a language with which it has precious little in common.

— BILL BRYSON, *Mother Tongue: English and How it Got That Way* (1990)

If English is not your first language then writing mathematics is doubly difficult, since you have to grapple with grammar and vocabulary as you try to express your thoughts. In this chapter I offer a few basic suggestions and give some references for further reading. The problems encountered by a non-native writer of English depend very much on the writer's first language. I concentrate on general advice that is not specific to any particular language.

5.1. Thinking in English

Try to think in English and construct sentences in English. If you compose in your own language and then translate you are more likely to produce prose that is not idiomatic (that is, does not agree with standard usage). Here are some examples of non-idiomatic phrases or sentences:

> ▷ *Capable to do.* Although this is logically correct, convention requires that we say "capable of doing".

> ▷ *We have the possibility to obtain an asymptotic series for the solution.* We do not normally say "possibility to". Better is *It is possible to obtain* . . . (passive voice), or, shorter, *We can obtain* . . . .

> ▷ *This result was proved already in* [5]. *Already* should be deleted or replaced by *earlier* or *previously*. Alternative: *This result has already been proved in* [5].

> ▷ *The solution has been known since ten years.* This type of construction occurs in those European languages in which one word serves for both *for* and *since*. In this example, *since* should be replaced by *for*.

> ▷ *This approach permits to exploit the convexity of f.* The phrase should be *permits us to exploit* (active voice) or *permits exploitation of* (passive voice). Or, depending on the context, it may be acceptable to shorten the sentence to *This approach exploits the convexity of f*.

> ▷ *To our experience* . . . . The correct phrase is *In our experience* . . . .

> ▷ *We invoke again Theorem* 4.1. This sentence is correct, but does not sound quite right to a native speaker of English. Better is *We invoke Theorem* 4.1 *again* or *Once again we invoke Theorem* 4.1.

▷ *The method is easy to use and performant.* There is no
word *performant* (even though the verb *converge*, for example, produces the adjective *convergent*). "Performs well"
is probably the intended meaning.

▷ *In the next section we give some informations about the
network of processor used.* Here, the problem is with the
plurality of nouns: it should be *information* (an uncountable noun) and *processors*.

Part of the process of thinking in English is to write your research notes
in English from the start and to annotate the books and papers you study
in English.

5.2. Reading and Analysing Other Papers

Read as many well-written papers in your field as you can. Ask a friend or
colleague for recommendations. Analyse the following aspects.

Vocabulary. What kinds of words occur frequently in technical writing?
Look up words that you don't know in a dictionary. For technical terms you
may need to use a mathematical or scientific dictionary or encyclopedia.
Note the range of applicability of particular words.

Synonyms. Notice different ways of saying the same thing and different
ways of ordering sentences. Use what you learn to avoid monotony in your
writing.

Collocations. These are groups of words that commonly appear together. For example, feeble and fragile are synonyms for weak, but weak
has more meanings, and while we readily say "weak bound" we never say
"feeble bound" or "fragile bound". As another example, we say "uniquely
determined" or "uniquely specified" but not "uniquely fixed" or "uniquely
decided". Build up a list of the collocations you find in mathematical writing.

Idioms. Idioms are expressions whose meanings cannot be deduced from
the words alone, but are established by usage. Here are some examples of
idioms that are sometimes found in technical writing (and more commonly
in speaking), with the idiom in the left column and a definition in the right
column.

By and large.	Taking everything into account.
End up.	Reach a state eventually.
In that.	In so far as.
It goes without saying.	Something so obvious that it needn't be said.
On the other hand.	From the other point of view.
On the whole.	In general, ignoring minor details.
Over and above.	In addition to.
Rule of thumb.	A rule based on experience and estimation rather than precise calculation.
Start from scratch.	To start from the beginning with no help.
Trial and error.	Attempting to achieve a goal by trying different possibilities to find one that works.

Errors in the use of idioms tend to be very conspicuous. It is good advice to avoid idioms until you are sure how to use them correctly.

5.3. Distinctions

Satisfy, Verify. These words can be difficult to distinguish for some nonnative speakers, who often incorrectly use *verify* for *satisfy*. In mathematics, *verify* means to establish the truth of a statement or equation, and is a synonym for check; it is the mathematician who verifies. On the other hand, a quantity *satisfies* an equation if it makes it true. Thus we write "We now verify that x is a global maximizing point of f" but "We have to show that x satisfies the sufficient conditions for a global maximizer."

5.4. Articles

Some languages either do not have articles (words such as "the", "a" and "an") or use them in a different way than in English, so it is difficult for speakers of these languages to use articles correctly in English. The rules of article use are complicated. Swan [266] explains them and identifies two of the most important.

(1) Do not use *the* (with plural or uncountable nouns) to talk about things in general. Examples: "Mathematics is interesting" (not "The mathematics is interesting"); "Indefinite integrals do not always have closed form solutions" (not "The indefinite integrals do not always have the closed form solutions").

(2) Do not use singular countable nouns without articles. Examples: "the derivative is", "a derivative is", but not "derivative is".

In certain circumstances an article is optional. The sentences "A matrix with the property (3.2) is well conditioned" and "A matrix with property

(3.2) is well conditioned" are both correct.

Mistakes in the use of articles are undesirable, but they do not usually obscure the meaning of a sentence.

5.5. Ordinal Numbers

Here are examples of how to describe the position of a term in a sequence relative to a variable k:

kth, $(k + 1)$st, $(k + 2)$nd, $(k + 3)$rd, $(k + 4)$th, ...
(zero*th*, fir*st*, seco*nd*, thi*rd*, four*th*, ...)

Generally, to describe the term in position $k \pm i$ for a constant i, you append to $(k \pm i)$ the ending of the ordinal number for position i (th, st, or nd), which can be found in a dictionary or book of grammar.

5.6. Negatives

A double negative results when two words with negative meanings are used together. Double negatives are commonly used in some languages (for example, French and Spanish) as a way of expressing a single negative idea. In English, however, two negative words combine to give a positive meaning. Double negatives are sometimes used for special effect, but they should be avoided in technical writing. Examples of double negatives are

▷ We do not know nothing about the location of the roots. (Literally means "We know something." Replace by *We know nothing* or *We do not know anything.*)

▷ The method hasn't never failed to work in our experience. (Literally means "The method has failed." Replace *never* by *ever* or *hasn't* by *has.*)

5.7. Constructions

Certain constructions are common in mathematical writing. You may find it helpful to make a list for reference, beginning with the following entries. The left column contains constructions, and the right column examples.

Let ... be	Let f be a continuous function.
If ... then	If $\alpha > -1$ then the integral exists.
Suppose (that) ... is/are	Suppose g is differentiable.
	Suppose that A and B have no eigenvalue in common.
We define ... to be	We define a problem to be stable if ...
It is easy to see/show that	It is easy to show that the error decays as t increases.
From ... we have	From (5.2) we have the inequality ...
By substituting ... into ... we obtain	By substituting (1.9) into (7.3) we obtain ...
A lower bound for	A lower bound for h can be obtained from ...
Without loss of generality	Without loss of generality we can assume that $x > 0$.

5.8. Connecting Words and Phrases

In this section I give examples of the use of words and phrases that connect statements. Most of the examples are followed by comments on the degree of emphasis; note, however, that the emphasis imparted sometimes depends on the context in which the word or phrase appears, so extrapolation from these examples should be done with care. Mastering these connectives, and the differences between them, is an important part of learning to write technical English. This section is loosely based on [78, pp. 191–194].

Combinations

Statement a: Direct methods are used to solve linear systems.
Statement b: Iterative methods are used to solve linear systems.

and	Direct methods *and* iterative methods are used to solve linear systems.
	(No emphasis on either type of method.)
both	*Both* direct and iterative methods are used to solve linear systems.
	(Similar to *and*.)
also	Direct methods, and also iterative methods, are used to solve linear systems.
	Direct methods are used to solve linear systems, as also are iterative methods.
	(Slight emphasis on direct methods.)

as well as Direct methods, as well as iterative methods, are used to solve linear systems.
(Similar to *also*.)

not only ... but also Linear systems can be solved *not only* by direct methods *but also* by iterative methods.
(Emphasizes that there is more than one possibility.)

apart from/in addition to *Apart from* (*in addition to*) direct methods, iterative methods are used to solve linear systems.
(Emphasizes that there is more than one type of method; slightly more emphasis on direct methods than *also* but less than *not only ... but also*.)

moreover/furthermore The name of Gauss is attached to the most well-known method for solving linear systems, Gaussian elimination. *Moreover* (*furthermore*), a popular iterative technique also takes his name: the Gauss–Seidel method.
(Stresses the statement after *moreover/furthermore*.)

Implications or Explanations

Statement a: The problem has a large condition number.
Statement b: The solution is sensitive to perturbations in the data.

as/because/since *As* (*because, since*) the problem has a large condition number, the solution is sensitive to perturbations in the data.
The solution is sensitive to perturbations in the data, *as* (*because, since*) the problem has a large condition number.

due to The sensitivity of the solution to perturbations in the data is *due to* the ill condition of the problem.
(More emphatic than *as*.)

in view of/owing to/on account of *In view of* (*owing to, on account of*)[7] the ill condition of the problem, the solution is sensitive to perturbations in the data.
(More emphatic than *as*.)

---

[7] *Due to* would be incorrect here; see §4.14.

given   *Given* the ill condition of the problem, the solution is necessarily sensitive to perturbations in the data.

(Inevitable result of the stated condition.)

it follows that   The problem has a large condition number. *It follows that* the solution is sensitive to perturbations in the data.

(Puts more emphasis on the first statement than *as*.)

consequently/therefore/thus The problem has a large condition number and *consequently* (*therefore, thus*) the solution is sensitive to perturbations in the data.

(Intermediate between *as* and *it follows that*. *Consequently* and *therefore* are preferable to *thus* at the beginning of a sentence.)

Modifications and Restrictions

Statement a: Runge-Kutta methods are widely used for solving non-stiff differential equations.

Statement b: For stiff differential equations, methods based on backward differentiation formulae (BDF) are preferred.

alternatively If the differential equations are non-stiff, Runge-Kutta methods can be used; *alternatively*, if the differential equations are stiff, BDF methods are preferred.

although   *Although* Runge-Kutta methods are widely used for non-stiff differential equations, BDF methods are preferred when the differential equations are stiff.

(More emphasis on BDF methods than *alternatively*.)

though   Runge-Kutta methods are widely used for non-stiff differential equations, *though* BDF methods are preferred when the differential equations are stiff.

(*Though* is weaker than *although*, and it tends to be used inside a sentence rather than at the beginning. In this example *though* could be replaced by *although*, which would give greater

	emphasis to the BDF methods.)
but	If the differential equations are non-stiff, Runge-Kutta methods can be used, *but* if the differential equations are stiff, BDF methods are preferred.
	(Similar to *though.*)
whereas	BDF methods are used for stiff differential equations, *whereas* Runge-Kutta methods are used for non-stiff equations.
by contrast	Runge-Kutta methods are widely used for non-stiff differential equations. *By contrast*, for stiff equations BDF methods are the methods of choice.
except	*Except* for stiff differential equations, for which BDF methods are preferred, Runge-Kutta methods are widely used.
	(Clearly defined limitation or restriction.)
however[8]/on the other hand	Runge-Kutta methods are widely used for solving non-stiff differential equations. *However (on the other hand)*, for stiff differential equations BDF methods are preferred. (Note that *however* and *on the other hand* are not always interchangeable. *On the other hand* is applicable only when there are two possibilities, corresponding to our two hands! This example is similar to *although* and *though*, but it merely joins the two statements.)
nevertheless	BDF methods are much less well known than Runge-Kutta methods. *Nevertheless*, there is great demand for BDF codes.
	(The second statement is true even though the first statement is true. It would not be correct to replace *however* by *nevertheless* in the previous example, although these two words are sometimes interchangeable.)
despite/in spite of	*Despite (in spite of)* the stiff nature of the differential equations that arise in his chemical reaction problems, Professor X prefers to use his favourite Runge-Kutta code.

---

[8] *However* can be used with another meaning. "However Runge-Kutta methods are used" means "no matter how Runge-Kutta methods are used".

> (Even though his differential equations are stiff, Professor X uses his Runge-Kutta code.)

instead of/rather than For stiff differential equations we use BDF methods *instead of* (*rather than*) Runge-Kutta methods. *Instead of* using (*rather than* use) our usual Runge-Kutta code we turned to a BDF code because we thought the differential equations might be stiff.

> (*Rather than* and *instead of* are generally interchangeable but their meaning is different: *rather than* implies a conscious choice, whereas *instead of* merely states an alternative.)

Conditions

Statement a: The indefinite integral does not have a closed form solution.
Statement b: Numerical integration provides an approximation to the definite integral.

if *If* the integral cannot be evaluated in closed form, numerical integration should be used to obtain an approximation.

unless *Unless* the integral can be evaluated in closed form, numerical integration should be used to obtain an approximation.

> (Using the converse of the logical condition for *if*.)

whether or not *Whether or not* a closed form exists, numerical integration will always provide an approximation to the integral.

> (A numerical approximation can be obtained independent of whether a closed form exists for the integral.)

provided (providing) that *Provided* (*providing*) *that* a closed form solution does not exist, the student may resort to numerical integration.

> (More restrictive than *if*. The *that* following *provided* or *providing* can often be omitted, as it can in this sentence.)

Emphasis

Statement: Gaussian elimination with row interchanges does not break down; a zero pivot is a welcome event for it signals that the column is already in triangular form and so no operations need be performed during that step of the reduction.

indeed   Gaussian elimination with row interchanges does not break down; *indeed,* if a zero pivot occurs it signals that the column is already in triangular form and so no operations need be performed during that step of the reduction.

(Introducing a further observation that builds on the first one.)

actually A zero pivot in Gaussian elimination with row interchanges is *actually* a welcome event, for it signals that the column is already in triangular form and so no operations need be performed on that step of the reduction.

(Here, *actually* emphasizes the truth of the statement. It could be replaced by *indeed,* but in the first example *indeed* could not be replaced by *actually;* this is because both words have several slightly different meanings.)

clearly A zero pivot in Gaussian elimination with row interchanges signals that the column is already in triangular form and that no operations need be performed on that step of the reduction, so the method *clearly* (*obviously, certainly*) cannot break down.

5.9. Spelling

Many English words have alternative spellings. These alternatives fall broadly into two classes. The first class contains those words that are spelled differently in British and American English. Table 5.1 illustrates some of the main differences. A special case worthy of mention is the informal abbreviation for mathematics: maths (British English), math (American English). Obviously, you should use British spellings or American spellings but not a mixture in the same document.

The second class of words is those that have alternative spellings in British English.

Table 5.1. British versus American spelling.

British spelling	American spelling
behaviour	behavior
catalogue	catalog or catalogue
centre	center
defence	defense
grey	gray
manoeuvre	maneuver
marvellous	marvelous
modelled	modeled
modelling	modeling
skilful	skillful
speciality	specialty

embed, imbed

learnt, learned. The past tense of the verb *learn*. The advantage of *learnt* is that it avoids confusion with the other meaning of *learned*, which is "having much knowledge acquired by study" (and which is pronounced differently from the first meaning).

spelt, spelled. The past tense of the verb *spell*.

In the last two examples, the -ed ending is the form used in American English. Other examples are

acknowledgement	acknowledgment
benefited	benefitted
encyclopaedia	encyclopedia
focused	focussed
judgement	judgment

There is a host of words, mostly verbs, that can take an -ise or -ize ending. Examples are *criticize* and *minimize*. The -ize ending is used in American English, while the -ise ending tends to be preferred in British English (even though *The Concise Oxford Dictionary* gives prominence to the -ize form). A number of words, including the following ones, take only an -ise ending:

advise, arise, comprise, compromise, concise, devise, disguise, exercise, expertise, likewise, otherwise, precise, premise, reprise, revise, supervise, surmise, surprise, treatise.

Verbs ending in -yse (such as *analyse* and *catalyse*) are spelt -yse in British English and -yze in American English.

Several plurals have alternative spellings:

appendices	appendixes
formulae	formulas
indices	indexes
lemmata	lemmas
vertices	vertexes

There are also pairs of words that have different meanings in British English but for which one of the pair is often (or always) used with both meanings in American English. Examples: ensure and insure (insure is frequently used for ensure in American English), practice and practise (see §4.14). These differences are subtle, and mastering them is not vital to producing clear and effective prose (native speakers also find them confusing). For a good explanation of the reasons for the often haphazard spelling of English, and the reasons for the differences between British and American spelling, I recommend the book by Bryson [42].

Finally, it is important to be aware of words that have very similar spellings or pronunciations, but different meanings. Examples:

accept (agree to receive)	except (but, excluding)
adapt (modify)	adopt (take up, accept)
advice (noun)	advise (verb)
affect (verb: influence)	effect (noun: result, verb: bring about)
	(see also §4.14)
complement (the rest)	compliment (flattering remark)
dependant (noun)	dependent (adjective)
device (noun: scheme)	devise (verb: invent)
discreet (careful)	discrete (not continuous)
precede (go before)	proceed (continue)
principal (main)	principle (rule)
sign	sine (trigonometric function)
stationary (fixed)	stationery (materials for writing)

5.10. Keeping It Simple

The best way to avoid making errors is to keep your writing simple. Use short words and sentences and avoid complicated constructions. Such writing may not be elegant, but it is likely to be unambiguous and readily understood. As your knowledge of English improves you can be more ambitious with your sentence structure and vocabulary. Here is an example giving simple and complicated versions of the same paragraph.

Simple: We note that if the transformation matrix H_p has large elements then A_p is likely to have much larger elements than A_{p-1}. Therefore we can expect large rounding errors in the pth stage, in which case the matrix F_p will have large norm.

Complicated: This conclusion is reinforced if we observe that a transformation with a matrix H_p having large elements is likely to give an A_p with much larger elements than A_{p-1}, and the rounding errors in this stage are therefore likely to be comparatively large, leading to an F_p with a large norm.

5.11. Using a Dictionary

In addition to a bilingual dictionary, you should buy a monolingual English dictionary and a thesaurus (preferably hardback ones, as you will be using them a lot!). Most bilingual dictionaries do not provide enough detail or wide enough coverage for the writer of scientific English. Also, using a monolingual dictionary will help you to think in English. Instead of, or in addition to, a general-purpose dictionary, you may want to acquire a dictionary written for advanced learners of English, of which there are several. These dictionaries have several notable features. They

- describe a core vocabulary of contemporary English;

- use simple language in their definitions (the *Longman's Dictionary* mentioned below uses a special defining vocabulary of 2000 words);

- give guidance on grammar, style and usage;

- provide examples illustrating typical contexts;

- show pronunciation;

- in some cases, show allowable places to divide a word when it must be split at the end of a line. (This information can also be found in special-purpose dictionaries of spelling and word division.)

Three such dictionaries described as "outstanding" by Quirk and Stein [232] are the *Longman Dictionary of Contemporary English* [181], *Collins Cobuild English Dictionary* [59] and the *Oxford Advanced Learner's Dictionary of Current English* [212]. Another dictionary that you may find useful is *Collins Plain English Dictionary* [61], which has very easy to read definitions.

Whereas a dictionary provides meanings for a word, a thesaurus lists alternatives for it that have approximately the same meaning (synonyms).

Most thesauruses are arranged alphabetically, like a dictionary. In preparation for using your dictionary and thesaurus you should learn the abbreviations they use (these are usually listed at the front) and make sure you understand the grammatical terms noun, adjective, adverb and (transitive or intransitive) verb.

When you are looking for the right word to express an idea, pick whatever words you already know (or that you find in a bilingual dictionary) and look them up in the thesaurus. Then look up in the dictionary the definitions of the synonyms you find and try to decide which is the most appropriate word. This may take some time. It is worth making a note to summarize the search you conducted, as you may later want to retrace your steps.

Watch out for "false friends"—two words in different languages that are very similar but have different meanings. Examples:

> French: *actuellement* = at present, currently. Cf. actually.
> Italian: *eventuale* = possible. Cf. eventual.
> German: *bekommen* = get, receive. Cf. become.

Part of the task of choosing a word is choosing the correct part of speech. Suppose you write, as a first attempt,

> *The interested Jacobian matrices are those with large, positive dominate eigenvalues.*

The two words most likely to be wrong are *interested* and *dominate*, as they can take several different forms. *The Concise Oxford Dictionary* (COD) says dominate is a verb, one meaning of which is "have a commanding influence on". The word we are looking for is an adjective, as it describes a property of eigenvalues. The previous dictionary entry is *dominant*, an adjective meaning most influential or prevailing. This is the correct word. The word *interested* is the correct part of speech: it is an adjective, as it should be since it describes the Jacobian matrices. The COD defines interested as meaning "having a private interest; not impartial or disinterested". It is therefore incorrect to talk about an *interested Jacobian matrix*. The word we require is *interesting*, another adjective meaning "causing curiosity; holding the attention". This example indicates how useful a dictionary can be if you are unsure of vocabulary and grammar. It is a good idea to look up in the dictionary every nontrivial word you write, to check spelling, meaning and part of speech. Do this after you have written a paragraph or section, so as not to interrupt your train of thought.

Using analogy to reason about English vocabulary works often, but not always. The noun *indication* is related to the verb *indicate*, and the pattern (noun, verb) = (-ion, -ate) is common. There are exceptions, however, and the one that most often causes trouble to the mathematical writer is the pair *perturbation* and *perturb*—there is no verb *perturbate*.

You will not be able to write perfect English simply by using a dictionary, because even the learner's dictionaries cannot tell you everything you need to know. The best way to learn the subtler aspects of English is to ask someone more fluent in English than yourself to comment on what you have written. If you are not used to writing in English it is almost obligatory to obtain such advice before you submit a paper for publication. Even better is to have a fluent speaker as a co-author. Make sure that you learn from the corrections and suggestions you receive, and keep a note of common mistakes, so as to avoid them.

5.12. Punctuation

There can be differences in punctuation between one language and another. In English, a decimal point separates the integer part of a number from its fractional part ($\pi \approx 3.141$) but in some European languages a comma is used instead ($\pi \approx 3{,}141$). In English, a comma is used to indicate thousands (2,135), but in French, until quite recently, a full stop was used for this purpose (2.135). (Note that a full stop in British English is a period in American English).

Some examples of different sentence punctuation follow, where _____ denotes a sentence or phrase in the given language.

Question: English: _____?
              Greek: _____;
              (romanized) Japanese: _____ ka
              Spanish: ¿_____?

Quotation: English: "_____"
              French, Italian, Spanish: «_____»
              German: „_____" or »_____«
              Swedish: "_____"

Semicolon: English: _____;
              Greek: _____· (but now little used)

5.13. Computer Aids

If you are writing on a computer make regular use of a spelling checker. As well as using a spelling checker to check text already typed, you may also be able to use it to complete or correct a spelling you are unsure of before you put the word into your document. If a style checker is available and you find it helpful, use it too. See §§14.5 and 14.6 for more about spelling checkers and style checkers.

5.14. English Language Qualifications

Two examinations are widely accepted for fulfilling university entrance requirements.

- The Test of English as a Foreign Language (TOEFL) is administered by the Educational Testing Service (required grade 500–580). This test is universally accepted in the US and is becoming more widely accepted in the UK.

- The International English Language Testing System (IELTS) is jointly managed by the British Council, the University of Cambridge Local Examinations Syndicate (UCLES) and the International Development Program of Australian Universities and Colleges (IDP). It is administered by UCLES. This test is widely accepted in the UK and Australia (minimum scores for undergraduate or postgraduate entry range from 4.0 to 7.5) and by an increasing number of institutions in the US.

Ideally, you will have attained these standards before you attempt to write research papers in English. English language courses for non-native speakers are available at many institutions.

5.15. Further Reading

A useful guide on how to use an English dictionary is the booklet by Underhill [282]. It is aimed particularly at users of learner's dictionaries from Oxford University Press and contains many exercises.

Swan's *Practical English Usage* [266] is an invaluable guide for the intermediate or advanced learner of English. It is arranged in dictionary form and covers vocabulary, idiom, grammar, style, pronunciation and spelling. It lists many common examples of incorrect usage, explains differences between British and American English, and distinguishes between formal and informal usage. The *Longman Dictionary of Common Errors* [136] aims to help the reader "avoid or correct those errors most frequently made by foreign learners of English".

Herbert's *The Structure of Technical English* [138] is aimed at students who have completed about six years of learning English and describes "the special structures and linguistic conventions of the English used in technical and scientific writing". This book has been described as pioneering, for being the first textbook in the area now known as English for Specific Purposes (ESP) [265]. Similar books are Swales's *Writing Scientific English* [264] and Ewer and Latorre's *A Course in Basic Scientific English* [78]; the latter contains a useful "dictionary of basic scientific English".

Trzeciak's *Writing Mathematical Papers in English* [274] is a cookbook of phrases used in mathematical writing, ready for adaptation by the reader, and also offers advice on English grammar, using mathematically oriented illustrations.

In *Communicating in Science* [36] Booth offers a short chapter titled "Addressed to those for whom English is a foreign language," which includes examples of words that are often used incorrectly.

You may be able to find books on English writing that are designed for readers with your particular linguistic background. For example, I am aware of two excellent books for Swedish speakers. *The Writing Process* [31] is by members of a research group in the English Department at the University of Stockholm and is based on ideas of the Bay Area Writing Project at Berkeley, a widely used programme for improving writing in American schools and colleges. *Using Numbers in English*, by MacQueen [190], explains many aspects of using numbers, including how to say in words expressions such as "1.06", "2115", and "the 1800s"; various uses of "one" and "once"; numbers as adjectives or prepositions; and percentages.

Chapter 6
Writing a Paper

*The title of a book is its special tag, its distinguishing label.
The choice of words is limited only by the number of words in the English
language, yet how few of them show up in titles.*
— ELSIE MYERS STAINTON, *A Bag for Editors* (1977)

*Go straight to the point, rather than begin with an
historical reference or resounding banality
('There is much research interest at present in the
biochemistry of memory').*
— BERNARD DIXON, *Sciwrite* (1973)

*To write a reference, you must have
the work you're referring to in front of you.
Do not rely on your memory. Do not rely on your memory.
Just in case the idea ever occurred to you, do not rely on your memory.*
— MARY-CLAIRE VAN LEUNEN, *A Handbook for Scholars* (1992)

*It was said of Jordan's writings that
if he had 4 things on the same footing
as (a, b, c, d) they would appear as
$a, M'_3, \epsilon_2, \Pi''_{1,2}$.*
— J. E. LITTLEWOOD, *Littlewood's Miscellany*[9] (1986)

*May all writers learn the art (it is not easy) of
preparing an abstract containing the
essential information* in their compositions.
— KENNETH K. LANDES, *A Scrutiny of the Abstract. II* (1966)

---

[9]See [34].

We write scientific papers to communicate our ideas and discoveries. Our papers compete for the readers' attention in journals, conference proceedings and other outlets for scholarly writing. If we can produce well-organized papers that are expressed in clear, concise English, and that convey our enthusiasm and are accessible to people outside our particular speciality, then our papers stand a better than average chance of being read and being referenced. It is generally accepted that the standard of scientific writing is not high[10] [63], [71], [191], [299], so a well-written paper will stand out from the crowd.

In this chapter I examine issues specific to writing a paper, as opposed to the general principles discussed in the previous chapters. Much of the chapter is applicable to writing a thesis (see also Chapter 9), book, or review. After considering the general issues of audience and organization I explore the building blocks of a research paper, from the title to the reference list.

6.1. Audience

Your first task in writing a paper is to determine the audience. You need to identify a typical reader and decide the breadth of your intended readership. Your paper might be written for a mathematics research journal, an undergraduate mathematics magazine, or a book for pre-university students. The formality of the prose and mathematical developments will need to be different in each of these cases. For research journals, at one extreme, you might be writing a very technical paper that builds upon earlier work in a difficult area, and you might be addressing yourself only to experts in that area. In this case it may not be necessary to give much motivation, to put the work in context, or to give a thorough summary and explanation of previous results. At the other extreme, as when you are writing a survey paper, you do need to motivate the topic, relate it to other areas, and explain and unify the work you are surveying. The requirements set forth in the "Guidelines for Authors" for the journal *SIAM Review* (prior to 1998) are even more specific:

> In their introductory sections, all papers must be accessible to the full breadth of SIAM's membership through the motivation, formulation, and exposition of basic ideas. The importance and intellectual excitement of the subject of the paper must

---

[10]In 1985, the editor of the British Medical Journal said "The 5000 or so articles we see at the BMJ every year are mostly dreadfully written, with numerous faults in English and overall construction" [180, p. 232].

be plainly evident to the reader. Abstraction and specialization must illuminate rather than obscure. The primary threads of the intellectual fabric of the paper can never be hidden by jargon, notation, or technical detail.

The goals described in the last three sentences are worth striving for, whatever your audience.

The audience will determine the particular slant of your paper. A paper about Toeplitz matrices for engineers would normally phrase properties and results in terms of the physical problems in which these matrices arise, whereas for an audience of pure mathematicians it would be acceptable to consider the matrices in isolation from the application.

The language you use will depend on the audience. For example, whereas for linear algebraists I would write about the least squares problem $\min_x \|Ax - b\|_2$ with its least squares solution x, for statisticians I would translate this to the linear regression problem $\min_b \|Xb - y\|_2$ and the least squares estimate $\widehat{b}$ of the regression coefficients. Failing to use the notation accepted in a given field can cause confusion and can make your work impenetrable to the intended readership.

A good question to ask yourself is why a member of your intended audience should want to read your paper. If the paper is well focused you will find it easy to answer this question. If you cannot find an answer, consider altering the aims of the paper, or doing further work before continuing with the writing.

Whatever your audience, it is worth keeping in mind the words of Ivars Peterson, the editor of *Science News* [223]:

> The format of most journal papers seems to conspire against the broad communication of new mathematical ideas .... The titles, abstracts and introductions of many mathematical papers say: "Outsiders keep out! This is of interest only to those few already in the know."

With a little effort it should be possible to make your work at least partially understandable to non-experts.

6.2. Organization and Structure

At an early point in the writing of your paper you need to think about its high-level organization. It is a good idea to rank your contributions, to identify the most important. This will help you to decide where to put the emphasis and how to present the work, and it will also help you to write the title and abstract.

At the outset you should have some idea of the length of the paper. The larger it is the more important it is that it be well organized. However, it can be harder to write a good short paper than a good long one, for it is difficult to be simultaneously thorough, lucid and concise.

You need to decide in what order to treat topics and how to present results. You should aim to minimize the length by avoiding repetition; to obtain general results that provide others as special cases (even if the latter are more interesting); and to emphasize similarities and differences between separate analyses.

In addition to considering the first-time reader, you must think about the reader who returns to the paper some time after first studying it. This reader will want to skim through the paper to check particular details. Ideally, then, your paper should not only be easy to read through from beginning to end, but should also function as a reference document, with key definitions, equations and results clearly displayed and easy to find.

6.3. Title

According to Kerkut [150], for every person who reads the whole text of a scientific paper, five hundred read only the title. This statistic emphasizes the importance of the title. The title should give a terse description of the content, to help someone carefully scanning a journal contents page or a reference list decide whether to read the abstract or the paper itself. Ideally, it should also be catchy enough to attract the attention of a browser. Achieving a balance between these two aims is the key to writing an effective title. Kelley [147] mentions that he published an abstract with the title "A Decomposition of Compact Continua and Related Results on Fixed Sets under Continuous Mappings". After Paul Halmos suggested to him that it is not a good idea to put the whole paper into the title, he changed it to "Simple Links and Fixed Sets" for the published paper.

A note on the interpolation problem is too vague a title: what is the breakthrough heralded by *A note*, and which *interpolation problem* is under discussion? Similarly, *Approximation by cubic splines* is too vague (except for a thorough survey of the topic), since the approximation problem being addressed is not clear. Here are some examples of real titles, with my comments.

- Computing the eigenvalues and eigenvectors of symmetric arrowhead matrices [D. P. O'Leary and G. W. Stewart. *J. Comp. Phys.*, 90:497–505, 1990]. This is a lively and informative title. It is good to have action words in the title, such as *computing* or *estimating*.

- How and how not to check Gaussian quadrature formulae [Walter Gautschi. *BIT*, 23:209–216, 1983]. "How to" titles immediately arouse the reader's interest.

- Gaussian elimination is not optimal [V. Strassen. *Numer. Math.*, 13: 354–356, 1969]. If you can summarize your paper in a short sentence, that sentence may make an excellent title.

- How near is a stable matrix to an unstable matrix? [Charles F. Van Loan. In *Linear Algebra and Its Role in Systems Theory*, B. N. Datta, editor, volume 47 of *Contemporary Math.*, Amer. Math. Soc., 1985, pages 465–478]. A title that asks a question is direct and enticing.

- *Regression Diagnostics: Identifying Influential Data and Sources of Collinearity* [D. A. Belsley, E. Kuh, and R. E. Welsch. Wiley, New York, 1980]. This title has the classic form "general statement followed by colon and more specific information". Dillon [70] defines such a title to have "titular colonicity" and suggests that "To achieve scholarly publication, a research title should be divided by a colon into shorter and longer pre- and postcolonic clauses, respectively, the whole not to fall below a threshhold [sic] of 15–20 words minimum." This paper appears in the August (not the April) issue of *American Psychologist* and appears not to be a spoof.

- ALGOL 68 with fewer tears [Charles H. Lindsey. *Computer Journal*, 15:176–188, 1972]. This is an excellent title—it announces that the paper is about the programming language ALGOL 68 and whets the reader's appetite for a demystifying explanation. This paper has the distinction of being a syntactically valid ALGOL 68 program!

- Nineteen dubious ways to compute the exponential of a matrix [Cleve B. Moler and Charles F. Van Loan. *SIAM Review*, 20(4):801–836, 1978]. A classic paper in numerical analysis, with a memorable title.

- Can you count on your calculator? [W. Kahan and B. N. Parlett. Memorandum No. UCB/ERL M77/21, Electronics Research Laboratory, College of Engineering, University of California, Berkeley, April 1977]. Puns rarely make their way into titles, but this one is effective.

- Performing armchair roundoff analyses of statistical algorithms [Webb Miller. *Comm. Statist. Simulation Comput.*, B7(3):243–255, 1978]. An otherwise drab title transformed by the word *armchair*. It suggests a gentle approach to the error analysis.

- Tricks or treats with the Hilbert matrix [Man-Duen Choi. *Amer. Math. Monthly*, 90:301–312, 1983]. This is another attractive and imaginative title.

- Can one hear the shape of a drum? [Marc Kac. *Amer. Math. Monthly*, 73(4, Part II):1–23, 1966]. The meaning of this attention-grabbing title is explained on the third page of Kac's expository paper:

  > You can now see that the "drum" of my title is more like a tambourine (which is really a membrane) and that stripped of picturesque language the problem is whether we can determine Ω if we know all the eigenvalues of the eigenvalue problem

  $$\tfrac{1}{2} \nabla^2 U + \lambda U = 0 \quad \text{in} \quad \Omega,$$
  $$U = 0 \quad \text{on} \quad \Gamma.$$

  This paper won its author the Chauvenet Prize and the Lester R. Ford Award (see Appendix E). Kac's question is answered negatively in "One cannot hear the shape of a drum" [Carolyn Gordon, David L. Webb, and Scott Wolpert. *Bull. Amer. Math. Soc.*, 27(1):134–138, 1992], and his "hearing" terminology is used in "You can not hear the mass of a homology class" [Dennis DeTurck, Herman Gluck, Carolyn Gordon, and David Webb. *Comment. Math. Helvetici*, 64:589–617, 1989].

- The perfidious polynomial [James H. Wilkinson. In *Studies in Numerical Analysis*, G. H. Golub, editor, volume 24 of *Studies in Mathematics*, Mathematical Association of America, Washington, D.C., 1984, pages 1–28]. A delightful alliteration for a paper that explains why numerical computations with polynomials can be treacherous. This paper won its author the Chauvenet Prize (see Appendix E).

- Fingers or fists? (The choice of decimal or binary representation) [W. Buchholz. *Comm. ACM*, 2(12):3–11, 1959]. An analogy and an alliteration combine here to make an appealing title.

See Appendix E for many more examples of good titles.

A few years ago I submitted for publication a manuscript titled "Least Squares Approximation of a Symmetric Matrix". A referee objected to the title because it does not fully define the problem, so I changed it to "The Symmetric Procrustes Problem", which is even less informative if you are not familiar with Procrustes problems, but is perhaps more intriguing.

Unless the title is short it will have to be broken over two or more lines at the head of the paper (or on the front cover of a technical report). Rules of thumb are that a phrase should not be split between lines, a line should not start with a weak word such as a conjunction, and the lines should not differ too much in length. If the title is to be capitalized then all words except articles, short prepositions and conjunctions should be capitalized. Here are some examples, the first of each pair being preferable. The quotes at the beginning of each chapter provide further examples of the choice of line breaks.

On Real Matrices with
Positive Definite Symmetric Component

On Real Matrices with Positive
Definite Symmetric Component

An Iteration Method for the
Solution of the Eigenvalue Problem of
Linear Differential and Integral Operators

An Iteration Method for the Solution of
the Eigenvalue Problem of
Linear Differential and Integral Operators

Numerically Stable Parallel Algorithms
for Interpolation

Numerically Stable Parallel Algorithms for
Interpolation

In 1851 Sylvester published a paper with the title "Explanation of the Coincidence of a Theorem Given by Mr Sylvester in the December Number of This Journal, with One Stated by Professor Donkin in the June Number of the Same" [268]. Thankfully, titles are generally shorter nowadays.

6.4. Author List

In 1940, over 90% of papers reviewed in *Mathematical Reviews* had one author [120]. Today that figure is about 50%, showing that the proportion of jointly authored works in mathematics has increased greatly.

There are no hard and fast rules about the order in which the authors of a multiply authored paper are listed. Sometimes the person who did the greatest part of the work is listed first. Sometimes the academically senior person is listed first. In some disciplines and institutions the senior person

---

The Spotlight Factor

It is the custom in the theoretical computer science community to order authors alphabetically. In a tongue-in-cheek article, Tompa [273] defines the *spotlight factor* of the first author of a paper in which the k authors are listed alphabetically to be the probability that if $k-1$ coauthors are chosen independently at random they will all have surnames later in the alphabet than the first author. According to Tompa, the best (smallest) spotlight factor of 0.0255 in theoretical computer science belongs to Santoro, for his paper "Geometric containment and vector dominance" [Nicola Santoro, Jeffrey B. Sidney, Stuart J. Sidney, and Jorge Urrutia. *Theoretical Computer Science*, 53:345–352, 1987]. This value is calculated as

$$(1-.\text{santoro})^3 = \left(1 - \left(\frac{19}{27} + \frac{1}{27^2} + \frac{14}{27^3} + \frac{20}{27^4} + \frac{16}{27^5} + \frac{18}{27^6} + \frac{16}{27^7}\right)\right)^3,$$

where $a = 1, \ldots, z = 26$ and blanks or punctuation are represented by zero. By comparison, of those publications in the bibliography of this book, the best spotlight factors are the 0.1829 of O'Leary [211], the 0.2679 of Strunk [263] and the 0.3275 of Knuth [164].

(typically the director of a laboratory) is listed last. Perhaps most often, the authors are listed alphabetically, which is the practice I favour.

In their book *Computer Architecture* [137], Hennessy and Patterson adopt an unusual solution to the problem of deciding on author ordering: they vary the ordering in the book and in advertisements, even alternating the order when they reference the book! They comment that "This reflects the true collaborative nature of this book ... We could think of no fair way to reflect this genuine cooperation other than to hide in ambiguity—a practice that may help some authors but confuses librarians."

Being the first-named author is advantageous because the first name is easier to find in a reference list and the paper will be associated solely with that name in citations of the form "Smith et al. (1992)". Also, some citation services ignore all names after the first (see §14.3).

Make sure that you use precisely the same name on each of your publications. I declare myself as Nicholas J. Higham, but not N. J. Higham, N. Higham or Nick Higham. If you vary the name, your publications may not all be grouped together in bibliographic lists and indexes and there may be confusion over whether the different forms represent the same person.

Whether to spell out your first name(s) is a matter of personal preference. Chinese and Japanese authors need to decide whether to Westernize their name by putting the surname (family name) after the Christian (given) name, or to maintain the traditional ordering of surname first.

6.5. Date

Always date your work. If you give an unpublished paper to others they will want to know when it was written. You may not be able to distinguish different drafts if they are not dated. The date is usually placed on a line by itself after the author's name, or in a footnote. Spell the month out, rather than using the form "xx-yy-zz", since European and American authors interpret xx and yy in the opposite senses.

6.6. Abstract

The purpose of the abstract is to summarize the contents of the paper. It should do so in enough detail to enable the reader to decide whether to read the whole paper. The reader should not have to refer to the paper to understand the abstract.

Frequently, authors build an abstract from sentences in the first section of the paper. This is not advisable. The abstract is a mini-paper, and should be designed for its special purpose. This usually means writing the abstract from scratch once the paper is written. The shorter it is, the better, subject to the constraint of it being sufficiently informative. Most abstracts occupy one paragraph. Many mathematics journals state a maximum size for the abstract, usually between 200 and 300 words.

Some specific suggestions are as follows.

- Avoid mathematical equations in the abstract if possible, particularly displayed equations. One reason is that equations may cause difficulties for the review services that publish abstracts (though not those, such as *Mathematical Reviews*, that use TeX).

- Do not cite references by number in the abstract, since the list of references will not usually accompany the abstract in the review journals. If a paper must be mentioned, spell it out in full:

  An algorithm given by Boyd [*Linear Algebra and Appl.*, 9:95–101, 1974] is extended to mixed subordinate matrix norms.

- Try to make the abstract easy to understand for those whose first language is not English. Also, keep in mind that the abstract may be translated into another language, for a foreign review journal, for example.

- Some journals disallow the word *we* in abstracts, preferring the passive voice (usually necessitating *it*). If you are writing for such a journal it pays to adopt the required voice; while a copy editor will convert as necessary, the conversion may lessen the impact of your sentences.

- Obviously, the abstract should not make claims that are not justified in the paper. Yet this does happen—possibly when the abstract is written before the paper is complete and is not properly revised.

- The abstract should give some indication of the conclusions of the paper. An abstract that ends "A numerical comparison of these methods is presented" and does not mention the findings of the comparison is uninformative.

- Your abstract should lay claim to some new results, unless the paper is a survey. Otherwise, if you submit the paper to a research journal, you are making it easy for a referee to recommend rejection.

- Try not to start the abstract with the common but unnecessary phrases "In this paper" or "This paper". Some journals make this request in their instructions to authors.

The suggestions above are particularly relevant for an abstract that is submitted to a conference and appears in a conference programme. Such an abstract will be read and judged in isolation from the paper, so it is vital for it to create a strong impression in isolation.

An intriguing opening paragraph of an abstract is the one by Knuth (1979) in [158]:

ABSTRACT. Mathematics books and journals do not look as beautiful as they used to. It is not that their mathematical content is unsatisfactory, rather that the old and well-developed traditions of typesetting have become too expensive. Fortunately, it now appears that mathematics itself can be used to solve this problem.

If inspiration fails you, you could always use the following generic abstract from [246]:

ABSTRACT. After a crisp, cogent analysis of the problem, the author brilliantly cuts to the heart of the question with incisive simplifications. These soon reduce the original complex edifice to a [s]mouldering pile of dusty rubble.

6.7. Key Words and Subject Classifications

Some journals list key words supplied by the author, usually after the abstract. The number of key words is usually ten or less. Since the key words may be used in computer searches, you should try to anticipate words for which a reader might search and make them specific enough to give a good indication of the paper's content.

Some journals also require subject classifications. The AMS Mathematics Subject Classifications (1991) divide mathematics into 61 sections with numbers between 0 and 94, which are further divided into many subsections. For example, section 65 covers numerical analysis and has 106 subsections; subsection 65F05 covers direct methods for solving linear systems while subsection 65B10 covers summation of series. The classifications are listed in the *Annual Subject Index* of *Mathematical Reviews* and can be downloaded from the American Mathematical Society's e-MATH service (see §14.1).

Other classification schemes exist, such as the one for computer science from the journal *Computing Reviews*. The Computing Reviews Classification System (1987) is a four-level tree that has three numbered levels and an unnumbered level of descriptors. The top level consists of eleven nodes, denoted by letters A (General Literature) to K (Computing Milieux). An example of a category is

> G.1.3 [Numerical Analysis]: Numerical Linear Algebra—sparse
> and very large systems.

This specifies the Numerical Linear Algebra node of the Numerical Analysis area under G (Mathematics of Computation), and "sparse and very large systems" is one of several descriptors for G.1.3 listed in the definition of the classification scheme.

6.8. The Introduction

Perhaps the worst way to begin a paper is with a list of notation or definitions, such as *Let G be an abelian group and H be a subgroup of G,* or *Let $\mathcal{F}$ be the complex field $\mathcal{C}$ or the real field $\mathcal{R}$, and let $\mathcal{F}_{m \times n}$ be the linear space of all $m \times n$ matrices over $\mathcal{F}$. If $A \in \mathcal{F}_{m \times n}$ we use A^* to denote the conjugate transpose of A.* It can be argued that the first sentence should be the best in the paper—its job is to entice the uncommitted reader into reading the whole paper. A list of notation will not achieve this aim, but a clear, crisp and imaginative statement may. King [152] gives a slightly different specification for the first sentence: "The first sentence has a dual function: it must carry some essential information, particularly the problem under consideration, and at the same time gently translate the reader into the body of the article."

Ideally, an introduction is fairly short—say a few hundred words. It should define the problem, explain what the work attempts to do, and outline the plan of attack. Unless there is good reason not to do so, it is advisable to summarize the results achieved. Knowing the problem and the progress made on it, the reader can decide after reading the introduction whether to read the whole paper. You may want to leave the punch-line to the end, but in doing so you risk the reader losing interest before reaching it.

Here is an example of a compelling opening paragraph that strongly motivates the work. It is from Gautschi's paper "How and How Not to Check Gaussian Quadrature Formulae", mentioned on page 81.

> The preparation of this note was prompted by the appearance, in the chemistry literature, of a 16-digit table of a Gaussian quadrature formula for integration with measure $d\lambda(t) = \exp(-t^3/3)\, dt$ on $(0, \infty)$, a table, which we suspected is accurate to only 1–2 decimal digits. How does one go about convincing a chemist, or anybody else for that matter, that his Gaussian quadrature formula is seriously defective?

It can be appropriate to begin with a few definitions if they are needed to state the problem and begin the analysis. As an example, here is the beginning of the paper "Estimating the Largest Eigenvalue of a Positive Definite Matrix" by O'Leary, Stewart and Vandergraft [211].

> Let A be a positive definite matrix of order n with eigenvalues $\lambda_1 \geq \lambda_2 \geq \cdots \geq \lambda_n > 0$ corresponding to the orthonormal system of eigenvectors $x_1, x_2, \ldots, x_n$. In some applications, one must obtain an estimate of λ_1 without going to the expense of computing the complete eigensystem of A. A simple technique that is applicable to a variety of problems is the power method.

In the next four sentences the authors define the power method, state that the theory of the method is well understood, and note that convergence of the method can be hindered in two ways, which are then analysed. This introduction is effective because it defines the problem and motivates the analysis without a wasted word, and leads quickly into the heart of the paper. The introduction is not labelled as such: this is a three and a half page paper with no section headings.

A possible way to improve an introduction is to delete the first one or more sentences, which are often unimportant general statements. Try it! As an example, consider the following opening two sentences.

> Polynomials are widely used as approximating functions in many
> areas of mathematics and they can be expressed in various bases.
> We consider here how to choose the basis to minimize the error
> of evaluation in floating point arithmetic.

This opening is not a bad one, but the first sentence is general and unexciting. Under the reasonable assumption that the reader knows the importance of polynomials and need not be reminded, it is better to combine and shorten the two sentences and pose a question:

> In which basis should we express a polynomial to minimize the
> error of evaluation in floating point arithmetic?

Unless the paper is very short it is advisable to outline its organization towards the end of the introduction. One approach is to write, for each section, a sentence describing its contents. The outline can be introduced by a sentence such as "The outline of this paper is as follows" or "This paper is organized as follows." It is best not simply to list the section titles; instead, give a summary that could only be obtained by reading the sections.

Note that some journals prefer the symbol § to the word section when referring to specific sections: §2.1 is written instead of Section 2.1. The plural of § is §§: "see §§2.1 and 2.2."

6.9. Computational Experiments

Many papers describe computational experiments. These may be done for several reasons: to gain insight into a method, to compare competing methods, to verify theoretical predictions, to tune parameters in algorithms and codes, and to measure the performance of software. To achieve these aims, experiments must be carefully designed and executed. Many decisions have to be made, ranging from what will be the objectives of the experiments to how to measure performance and what test problems to use. One editor of a numerical analysis journal commented that his primary reason for rejecting papers is that the computational experiments are unsound; this underlines the importance of proper design and reporting of experiments.

When you report the results of a computational experiment you should give enough detail to enable the results to be interpreted and the experiment to be reproduced. In particular, where relevant you should state the machine precision (working accuracy) and the type of random numbers used (e.g., normal $(0,1)$ or uniform $(-1,1)$). If you wish to display the convergence of a sequence it is usually better to tabulate the errors rather than the values themselves. Error measures should be normalized. Thus, for an approximate solution $\widehat{x}$ to $Ax = b$, the relative residual $\|b - A\widehat{x}\|/(\|A\|\|\widehat{x}\| + \|b\|)$ is more meaningful than the scale-dependent quantity $\|b - A\widehat{x}\|$. If you are measuring the speed of a numerical algorithm it is important to show that the right answers are being produced (otherwise the algorithm "the answer is 42" is hard to beat).

You may also wish to state the programming language, the version of the compiler used, and the compiler options and optimizations that were selected, as these can all have a significant influence on run-times. In areas where attaining full precision is not the aim (such as the numerical solution of ordinary differential equations), it is appropriate to consider both speed and accuracy (e.g., by plotting cost versus requested accuracy).

One of the difficulties in designing experiments is finding good test problems—ones which reveal extremes of behaviour, cover a wide range of difficulty, are representative of practical problems, and (ideally) have known solutions. In many areas of computational mathematics good test problems have been identified, and several collections of such problems have been published. For example, collections are available in the areas of nonlinear optimization, linear programming, ordinary differential equations and partial differential equations. Several collections of test matrices are available and there is a book devoted entirely to test matrices. For references to test problem collections see [140].

In your conclusions you should make a clear distinction between objective statements and opinions and speculation. It is very tempting to extrapolate from results, but this is dangerous. As you analyse the results you may begin to formulate conclusions that are not fully supported by the data, perhaps because they were not anticipated when the experiments were designed. If so, further experimentation will be needed. When evaluating numerical algorithms I have found that it pays to print out every statistic that could conceivably be of interest; if I decide not to print out a residual or relative error, for example, I often find a need for it later on.

Further guidelines on how to report the results of computational experiments can be found in the journals *ACM Transactions on Mathematical Software* [65] and *Mathematical Programming* [146], and in an article by Bailey [12]. Perhaps the best advice is to read critically the presentation of experimental results in papers in your area of interest—and learn from them.

6.10. Tables

Many of the principles that apply to writing apply also to the construction of tables and graphs. However, some particular points should be considered when designing a table.

To maximize readability the table design should be as simple as possible. Repetition should be avoided; for example, units of measurement or descriptions common to each entry in a column should go in the column header. Compare Tables 6.1 and 6.2. It is best to minimize the number of rules in the table. Two busy examples are shown in Table 6.3 and Table 6.5. The simplified versions, Table 6.4 and Table 6.6, are surely more aesthetically pleasing. Table 6.3 is taken from [159, p. 366],[11] where it is given without any rules and is still perfectly readable.

It is easier to compare like quantities if they are arranged in columns rather than rows. Research reported by Hartley [132], [134] supports this fact, and Tables 6.7 and 6.8 provide illustration, Table 6.8 being the easier to read. The difference between row and column orientation is more pronounced in complex tables. Of course, the orientation may be determined by space considerations, as a horizontal orientation usually takes less space on the page. If a vertical table is too tall, but is narrow, it can be broken into two pieces side by side:

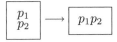

It is also helpful to put columns or rows that need to be compared next to each other.

Only essential information should be included in a table. Omit data whose presence cannot be justified and state only as many digits as are needed (this number is often surprisingly small). In particular, do not state numerical results to more significant figures than are known for the data. As an example, in Table 6.1 there is no need to quote the timings and speedups to six significant figures, so Table 6.2 gives just one decimal place. Note that there is justification for showing so many digits in Tables 6.3 and 6.4: $\pi(x)$ in this table is the number of primes less than or equal to x, and nearly all the digits of $\pi(10^9)$ are needed to show the error in Riemann's formula. Displaying the first one or two digits of the fractional parts of the approximations emphasizes that the approximations are not integers.

If you need to present a large amount of data in tabular form, consider displaying it in an appendix, to avoid cluttering the main text. You could give smaller tables in the text that summarize the data. Large sets of data are often better displayed as graphs, however, particularly if it is the trends rather than the numerical values that are of interest. Tufte [275,

---

[11]Donald E. Knuth, *The Art of Computer Programming*, vol. 2, ©1981 by Addison-Wesley Publishing Co. Reprinted by permission of Addison-Wesley Publishing Co., Inc., Reading, MA. The form of the original table is not reproduced exactly here.

Table 6.1. Timings for a parallel algorithm.

# processors	Time	Speedup
$p = 1$	28.352197 secs	—
$p = 4$	7.218812 secs	3.9275
$p = 8$	3.634951 secs	7.7999
$p = 16$	1.929347 secs	14.6952

Table 6.2. Timings for a parallel algorithm.

No. of processors	Time (secs)	Speedup
1	28.4	—
4	7.2	4.0
8	3.6	7.8
16	1.9	14.7

Table 6.3. Approximations to $\pi(x)$.

x	$\pi(x)$	$x/\ln x$	$L(x)$	Riemann's formula
10^3	168	144.8	176.6	168.36
10^6	78498	72382.4	78626.5	78527.40
10^9	50847534	48254942.4	50849233.9	50847455.43

Table 6.4. Approximations to $\pi(x)$.

x	$\pi(x)$	$x/\ln x$	$L(x)$	Riemann's formula
10^3	168	144.8	176.6	168.36
10^6	78498	72382.4	78626.5	78527.40
10^9	50847534	48254942.4	50849233.9	50847455.43

Table 6.5. Results for inverting a lower triangular matrix on a Cray 2.

	Mflops			
n	128	256	512	1024
Method 1 $(n_b = 1)$	95	162	231	283
Method 2 $(n_b = 1)$	114	211	289	330
k variant $(n_b = 1)$	114	157	178	191
Method 1B $(n_b = 64)$	125	246	348	405
Method 2C $(n_b = 64)$	129	269	378	428
k variant $(n_b = 64)$	148	263	344	383

Table 6.6. Mflop rates for inverting a lower triangular matrix on a Cray 2.

	n	128	256	512	1024
Unblocked:	Method 1	95	162	231	283
	Method 2	114	211	289	330
	k variant	114	157	178	191
Blocked:	Method 1B	125	246	348	405
$(n_b = 64)$	Method 2C	129	269	378	428
	k variant	148	263	344	383

Table 6.7. SI prefixes (10^{-1}–10^{12}). Row orientation.

Multiple	10^{12}	10^9	10^6	10^3	10^{-1}
Prefix	tera	giga	mega	kilo	deci
Symbol	T	G	M	K	d

Table 6.8. SI prefixes (10^{-1}–10^{12}). Column orientation.

Multiple	Prefix	Symbol
10^{12}	tera	T
10^9	giga	G
10^6	mega	M
10^3	kilo	K
10^{-1}	deci	d

p. 56] advises that tables are usually better than graphs at reporting small data sets of twenty numbers or less.

The caption should be informative and should not merely repeat information contained in the table. Notice the simplification obtained by moving the word "Mflops" from the table to the caption in Table 6.6.

Give a clear reference to the table at an appropriate place in the text— you cannot rely on the reader to refer to the table automatically. It is helpful if you explain the salient features of the table in words. The reader will appreciate this guidance, especially if the table contains a lot of data. However, you should not summarize the whole table—if you do, the table might as well be omitted.

Further Reading

The Chicago Manual of Style [58] devotes a whole chapter to tables and offers much useful advice. Another good reference is *A Manual for Writers* [278, Chap. 6]. Bentley [21, Chap. 10] gives a good example of how to redesign a table. References that discuss the preparation of graphs include Bentley [21, Chaps. 10, 11], Hartley [132], [134], MacGregor [187], [188], and Tufte [275], [276], [277].

6.11. Citations

The two main styles of citation in mathematics journals are by number (as used in this book) and by name and year, which is the Harvard system. Examples of the Harvard system are *These results agree with an existing study of Smith* (1990) and *These results agree with an existing study* (*Smith,* 1990). If more than one paper maps to Smith (1990), the papers are distinguished by appending a letter to the year: Smith (1990a), Smith (1990b), and so on. In the number-only system, the number is usually placed in square brackets, though some styles require it to be superscripted.

The main requirement is that a citation does not intrude upon a sentence. For example, *This method was found* [17] *to be unstable* is better written as *This method was found to be unstable* [17]. There are circumstances, however, where a citation has to be placed part-way through a sentence to convey the correct meaning. A good test for whether a citation is well placed is to see whether the sentence reads properly when the citation is deleted. The style of citation inevitably affects how you phrase sentences, so it is worth checking in advance what style is used by the journal in which you wish to publish. Knuth [164] explains that when his paper "Structured programming with go to statements" [*Computing Surveys,* 6:261–301, 1974] was reprinted in a book, he made numerous changes to make sentences read well with the citation style used in the book.

When you cite by number, it is good style to incorporate the author's name if the citation is more than just a passing one. As well as saving the reader the trouble of turning to the reference list to find out who you are referring to, this practice has an enlivening effect because of the human interest it introduces. Examples:

Let $A\Pi = QR$ be a QR factorization with column pivoting [10]. (Passing reference to a textbook for this standard factorization.) The rate of convergence is quadratic, as shown by Wilkinson [27]. (Instead of "as shown in [27]".)

The sentence "This question has been addressed by [5]" is logically incorrect and should be modified to "addressed by Jones [5]" or "addressed in [5]".

When you cite several references together it is best to arrange them so that the citation numbers are in increasing order, e.g., "several variations have been developed [2], [7], [13]." Ordering by year of publication serves no purpose when only citation numbers appear in the text. If you want to emphasize the historical progression it is better to add names and years: "variations have been developed by Smith (1974) [13], Hall (1981) [2], and Jones (1985) [7]."

It is important to be aware that the reference list says a lot about a paper. It helps to define the area in which the paper lies and may be used by a reader to judge whether the author is aware of previous work in the area. Some readers look at the reference list immediately after reading the title, and if the references do not look sufficiently familiar, interesting or comprehensive they may decide not to read further. Therefore it is desirable that your reference list contain at least a few of the key papers in the area in which you are writing. Papers should not, however, be cited just for effect. Each citation should serve a purpose within the paper. Note also that if you cite too often (say, for several consecutive sentences) you may give the impression that you lack confidence in what you are saying.

There are several conventions for handling multiple authors using the Harvard system. One such convention is as follows [45], [68, Chap. 12]. For one or two authors, both names are given (e.g., "see Golub and Van Loan (1989)"). If there are three authors, all three are listed in the first citation and subsequent citations replace the second and third names by "et al." (e.g., "see Knuth, Larrabee and Roberts (1989)," then "see Knuth et al. (1989)"). For four or more authors, all citations use the first author with "et al." These conventions can also be used when naming authors in conjunction with the numbered citation system.

If you make significant use of a result from another reference you should give some indication of the difficulty and depth of the result (and give the author's name). Otherwise, unless readers look up the reference, they will not be able to judge the weight of your contribution.

When you make reference to a specific detail from a book or long paper it helps the reader if your citation includes information that pinpoints the reference, such as a page, section, or theorem number.

For further details on the subtleties of citation consult van Leunen [283].

6.12. Conclusions

If there is a conclusions section (and not every paper needs one) it should not simply repeat earlier sections in the same words. It should offer another viewpoint, discuss limitations of the work, or give suggestions for further research. Often the conclusions are best worked into the introduction or the last section. It is not uncommon to see papers where the conclusions are largely sentences taken from the introduction, such as *We show that X's result can be extended to a larger class* ...; this practice is not recommended.

The conclusions section is a good place to mention further work: to outline open problems and directions for future research and to mention work in progress. Be wary of referring to your "forthcoming paper", for such papers can fail to materialize. A classic example of a justified reference to future work is the following quote from the famous paper[12] by Watson and Crick [290] (*Nature*, April 25, 1953) in which the double helix structure of DNA was proposed:

> It has not escaped our notice that the specific pairing we have postulated immediately suggests a possible copying mechanism for the genetic material.
>
> Full details of the structure, including the conditions assumed in building it, together with a set of co-ordinates for the atoms, will be published elsewhere.

The promise "will be published elsewhere" was fulfilled shortly afterwards, in the May 30, 1953 issue of *Nature*.[13]

6.13. Acknowledgements

Be sure to acknowledge any financial support for your work: grants, fellowships, studentships, sponsorship. A researcher might write "This re-

---

[12]For some comments by Crick on the writing of this and subsequent papers by Watson and Crick, see [64].

[13]Pyke [231] points out that the words "It has not escaped our notice that" can be removed.

search was supported by the National Science Foundation under grant DCR-1234567"; grant agencies like authors to be this specific. A Ph.D. student supported by the Engineering and Physical Sciences Research Council (UK) might write "This work was supported by an EPSRC Research Studentship." In SIAM journals this type of acknowledgement appears in a footnote on the first page. Other acknowledgements usually appear in a section titled *Acknowledgements* at the end of the paper. (An alternative spelling is *acknowledgments*.)

It is customary to thank anyone who read the manuscript in draft form and offered significant suggestions for improvement (as well as anyone who helped in the research), but not someone who was just doing his or her normal work in helping you (for example, a secretary). The often-used "I would like to thank" can be shortened to "I thank." Note that if you say "I thank X for pointing out an error in the proof of Theorem Y," you are saying that the proof is incorrect; "in an earlier proof" or "in an earlier attempted proof" is what you meant to say.

The concept of anonymous referee sometimes seems to confuse authors when they write acknowledgements. An anonymous referee should not be thanked, as is often the case, for *his* suggestions—it may be a *she*. One author wrote "I thank the anonymous referees, particularly Dr. J. R. Ockendon, for numerous suggestions and for the source of references." Another explained, not realising the two ways in which the sentence can be read, "I would like to thank the unknown referees for their valuable comments."

6.14. Appendix

An appendix contains information that is essential to the paper but does not fit comfortably into the body of the text. The most common use of an appendix is to present detailed analysis that would distract the reader if it were given at the point where the results of the analysis are needed. An appendix can also be used to give computer program listings or detailed numerical results. An appendix should not be used to squeeze inessential information into the paper (though this may be acceptable in a technical report or thesis).

6.15. Reference List

Preparing the reference list can be one of the most tedious aspects of writing a paper, although it is made much easier by appropriate software (see §13.3). The precise format in which references are presented varies among publishers and sometimes among different journals from the same publisher.

Here are four examples.

SIAM journals: J. H. WILKINSON, *Error analysis of floating-point computation*, Numer. Math., 2 (1960), pp. 319–340.

IMA journals: WILKINSON, J. H. 1960 Error analysis of floating-point computation. *Numer. Math.* **2**, 319–340.

Elsevier journals: J. H. Wilkinson, Error analysis of floating-point computation, *Numer. Math.* 2:319–340 (1960).

Springer-Verlag journals: Wilkinson, J. H. (1960): Error analysis of floating-point computation. Numer. Math. **2**, 319–340.

All journals that I am familiar with ask for the use of their own format but will accept other formats and copy edit them as necessary. All publishers have a minimum amount of information that they require for references, as defined in their instructions for authors. It is important to provide all the required information, whatever format you use for the references.

Here are some comments and suggestions on preparing reference lists. For further details I strongly recommend the book by van Leunen [283], but keep in mind that her recommendations may conflict with those of publishers in certain respects.

1. Do not rely on secondary sources to learn the contents of a reference or its bibliographic details—always check the original reference. In studies on the accuracy of citation, the percentage of references containing errors has been found to be as high as 50% [95]. A 1982 paper by Vieira and Messing in the journal *Gene* had been cited correctly 2,212 times up to 1988, but it had also been cited incorrectly 357 times under "Viera"; these errors led to the paper being placed too low in a list of most-cited papers [98]. In another well-documented case in the medical literature, the Czech title "O Úplavici" ("On Dysentery") of an 1887 paper in a Czech medical journal was taken by one writer to be the author's name, and the mistake propagated until it was finally exposed in 1938 [135], [239]. If a secondary source has to be used (perhaps because the reference is unavailable), it is advisable to append to the reference "cited in [ss]", where [ss] is the secondary source.

2. Always provide the full complement of initials of an author, as given in the paper or book you are referencing.

3. Some authors are inconsistent in the name they use in their papers, sometimes omitting a middle initial, for example. In such cases, my

preference is to use the author's full name in the reference list when it is known, to avoid ambiguities such as: Is A. Smith the same author as A. B. Smith?

4. Some sources contain typographical errors or nonstandard usage. Titles should be given unaltered. For example, the title "Van der Monde systems and numerical differentiation" [J. N. Lyness and C. B. Moler, *Numer. Math.*, 8:458–464, 1966] appears to be incorrect because the name is usually written Vandermonde, but it should not be altered (I have occasionally had to reinstate "Van der Monde" in my reference list after a copy editor has changed it). A typographical error in an author's name is rare, but not unknown. It seems reasonable to correct such an error, but to provide some indication of the correction that has been made, such as a note at the end of the reference.

5. Copy bibliographic information of a journal article from the journal pages, not the cover of the journal. The cover sometimes contains typographical errors and you cannot deduce the final page number of the article if the journal puts blank pages between articles or begins articles part-way down a page.

6. Electronic journals do not usually cause any difficulties in referencing, since it is in the journals' interests to make clear how papers should be referenced. For example, the journal *Electronic Transactions on Numerical Analysis* provides papers in PostScript form, and each paper has a clearly defined page range, volume and year; papers are therefore referenced just like those in a traditional journal. It may help readers if a URL for an electronic journal is appended to the reference, but the journal in which you are publishing may delete it to save space.

It is more difficult to decide how to reference email messages and unpublished documents or programs on the Web. The following suggestions are adapted from those in *Electronic Styles* [178]. I assume that an email address and a URL are both clearly identifiable as such by the @ and http, respectively, so I omit the descriptors "email" and "URL". There are so many different types of item on the Web that no referencing scheme can cover all possibilities.

(a) A publication available in print and online.

Nicholas J. Higham. The Test Matrix Toolbox for MAT-LAB (version 3.0). Numerical Analysis Report No. 276, Manchester Centre for Computational Mathematics, Manchester, England, Sept. 1995; also available from `ftp://ftp.ma.man.ac.uk/pub/narep/narep276.ps.gz`

(b) A publication available online only.

Melvin E. Page. A Brief Citation Guide for Internet Sources in History and the Humanities (Version 2.1), `http://h-net.msu.edu/~africa/citation.html`, 1996.

(c) A publication on CD-ROM.

A. G. Anderson, Immersed interface methods for the compressible equations. In Proceedings of the Eighth SIAM Conference on Parallel Processing for Scientific Computing (Minneapolis, MN, 1997), CD-ROM, Society for Industrial and Applied Mathematics, Philadelphia, PA, 1997.

(d) A piece of software.

Piet van Oostrum. LaTeX package `fancyhdr`. CTAN archive (e.g., `http://www.tex.ac.uk/tex-archive`), `macros/latex/contrib/supported/fancyhdr`.

(e) An item in a discussion list, digest or newsgroup.

David Hough. Random story. NA Digest, 89 (1), 1989. `na.help@na-net.ornl.gov`, `http://www.netlib.org/index.html`

(f) A standard email message. The title is taken from the `Subject:` line.

Desmond J. Higham (`aas96106@ccsun.strath.ac.uk`). Comments on your paper. Email message to Nicholas J. Higham (`higham@ma.man.ac.uk`), August 18, 1997.

(g) A forwarded email message.

Susan Ciambrano (`ciambran@siam.org`). Reader's comments on HWMS. (Original message A. Reader, Handbook of Writing.) Forwarded email message to Nicholas J. Higham (`higham@ma.man.ac.uk`), October 20, 1995.

7. When you reference a manuscript or technical report that is more than a few months old, check to see if it has appeared in a journal. An author will usually be happy to inform you of its status. If it has appeared, check whether the title has changed. The referees may have asked for a better title, or the copy editor may have added a hyphen, combined a hyphenated pair of words, or changed British spelling to American or vice versa.

8. Take care to respect letters and accented characters from other languages. Examples: Å, å, ß, é, ö, ø, Ø.

9. If you maintain a database of references (for example, in BibTeX format—see §13.3), it is worth recording full details of a reference, even if not all of them are needed for journal reference lists. For a journal article, record the part (issue) number as well as the volume number; this extra information can speed the process of looking up a reference, especially if the journal issues are unbound. For a technical report, the month of publication is useful to know.

10. Be sure you are using the correct journal name and watch out for journals that change their names. For example, the *SIAM Journal on Scientific and Statistical Computing* (1980–1992) became the *SIAM Journal on Scientific Computing* in 1993.

11. In book titles, van Leunen recommends that a colon be added if it is needed to separate a title from a subtitle, and an awkward comma or colon separating a title from a subtitle should be removed. Thus *On Writing Well An Informal Guide to Writing Nonfiction* (as copied from the title page of [304]) needs a colon added after *Well,* and the colon should be removed from *Interpolation Theory: 5*.

12. Van Leunen recommends simplifying the names of major publishers to the bare bones, so that *John Wiley & Sons* becomes *Wiley*, and *Penguin Books* becomes *Penguin*. She also recommends omitting the city for a major publisher; I usually include it because many journals require it. For obscure publishers it is best to give as complete an address as possible.

13. For a book, the International Standard Book Number (ISBN) is worth recording, as it can be used to search library and publishers' catalogues. (Note, though, that hardback and softback editions of a book usually have different ISBNs.) An ISBN consists of ten digits, arranged in four groups whose size can vary. The first group specifies the language group of the publisher ($0, 1$ = English speaking countries, 2 = French speaking, 3 = German speaking, etc.). The second group (2–7 digits) identifies the publisher (e.g., Oxford University Press is 19) and the third group (1–6 digits) identifies the particular title. The last digit is a checksum. If the ISBN is expressed as $d_1 d_2 \ldots d_{10}$ then

$$d_{10} = \lceil s/11 \rceil * 11 - s, \quad \text{where} \quad s = \sum_{i=1}^{9} (11 - i) d_i$$

($\lceil x \rceil$ denotes the smallest integer greater than or equal to x). A value $d_{10} = 10$ is written as "X". This book has the ISBN 0-89871-420-6: 89871 identifies SIAM as the publisher and 420 is the book's individual number. An International Standard Serial Number (ISSN) identifies a serial publication such as a journal, yearbook or institutional report. An ISSN has eight digits.

14. The date to quote for a book is the latest copyright date (excluding copyright renewals)—ignore dates of reprinting. Always state the edition number if it is not the first.

15. Make sure that every reference is actually cited in the paper. Some copy editors check this, as you may see from their pencilled marks on the manuscript when you receive the proofs.

16. Most mathematics journals require the reference list to be ordered alphabetically by author. Many science journals order by citation, so that the nth paper to be cited is nth in the reference list.

17. A list of standard abbreviations for mathematics journals can be found in *Mathematical Reviews* (see §14.3).

6.16. Specifics and Deprecated Practices

Capitalization

References to proper nouns should be capitalized: *See Theorem* 1.5, *the proof of Lemma* 3.4 *and the discussion in Section* 6. References to common nouns (generic objects) should not: *Next we prove the major theorems of the paper.*

Dangling Theorem

The term *dangling theorem* [147] (or *hanging theorem* [121]) refers to a construction such as the following one, where a theorem dangles or hangs from the end of a sentence.

> This result is proved in the following
> **Theorem 3.13**. If f is a twice continuously differentiable function ...

Halmos argues that while the practice can be defended, some readers dislike it, and it is not worth risking annoying them for the sake of avoiding the extra word *theorem*.

The following example does not strictly dangle, but is even more irritating.

> **5.1. Accuracy of the Computed Solution.** It depends on the machine precision and the conditioning of the problem.

Section headings stand alone and should not be taken as part of the text. The obvious solution

> **5.1. Accuracy of the Computed Solution.** The accuracy of the computed solution depends on the machine precision and the conditioning of the problem.

is inelegant in its repetition, but this could be avoided by rewriting the sentence or the title.

Footnotes

Footnotes are used sparingly nowadays in mathematical writing, and some journals do not allow them (see page 78 for an example of a footnote). It is bad practice to use them to squeeze more into a sentence than it can happily take. Their correct use is to add a note or comment that would deflect from the main message of the sentence. Donald W. Marquardt, the author of the 92nd most-cited paper in the Science Citation Index 1945–1988 [An algorithm for least-squares estimation of nonlinear parameters. *J. Soc. Indust. Appl. Math.*, 11(2):431–441, 1963], has stated that a critical part of the algorithm he proposed was described in a footnote and has sometimes been overlooked by people who have programmed the algorithm [97].

Numbering Mathematical Objects

Generally, you should number only those equations that are referenced within the text. This avoids the clutter of extraneous equation numbers and focuses the reader's attention on the important equations. Occasionally it is worth numbering key equations that are not referenced but which other authors might want to quote when citing your paper. Except in very short papers it is best to number equations by section rather than globally (equation (2.3) instead of equation (14)), for this makes referenced equations easier to find. The same applies to the numbering of theorems and other mathematical objects. Whether equation numbers appear on the left or the right of the page depends on the journal.

Two possible numbering sequences are illustrated by

Definition 1, Lemma 1, Theorem 1, Remark 1, Definition 2,
Lemma 2, ...
Definition 1, Lemma 2, Theorem 3, Remark 4, Lemma 5, ...

Opinions differ as to which is the best scheme. The last has the advantage
that it makes it easier to locate a particular numbered item, and the equation numbers themselves can even be included in the sequence for complete
uniformity. The disadvantage is that the scheme mixes structures of a different character, which makes it difficult to focus on one particular set of
structures (say, all the definitions); and on reading Remark 24 (say), the
reader may wonder how many previous remarks there have been. A compromise between the two schemes is to number all lemmas, theorems and
corollaries in one sequence, and definitions, remarks and so on in another.
Some typesetting systems control the numbering of mathematical objects
automatically. LaTeX does so, for example, and the numbering sequence for
definitions, lemmas, theorems, etc., can be specified by LaTeX commands.

Plagiarism

Plagiarism is the act of publishing borrowed ideas or words as though they
are your own. It is a major academic sin. In writing, if you copy a sentence
or more you should either place it in quotes and acknowledge the source via
a citation, or give an explicit reference such as "As Smith observed ...." In
the case of a theorem statement it is acceptable to copy it word for word if
you cite the source, but before copying it you should see whether you can
improve the wording or make it fit better into your notation and style.

Regarding when to quote and when to paraphrase, van Leunen [283]
advises "Quote what is memorable or questionable, strange or witty. Paraphrase the rest." When you wish to paraphrase, it is best to put the source
aside, wait a reasonable period, and then rewrite what you want to say in
your own words.

If you rework what you yourself have previously published without citing
the source, thus passing it off as new, that is self-plagiarism, which is no
less a sin than plagiarism.

Plagiarism has led to the downfall of many a career, in academia and
elsewhere. Some notable cases are described in Mallon's *Stolen Words* [192]
and LaFollette's *Stealing into Print* [169]. The former includes the ironic
news, quoted from the *New York Times* of June 6, 1980, that

> Stanford University said today it had learned that its teaching
> assistant's handbook section on plagiarism had been plagiarized
> by the University of Oregon. ... Oregon officials apologized and
> said they would revise their guidebook.

Fraud is another serious malpractice, though apparently and understandably rare in mathematical research. Numerous cases of scientific fraud through history are catalogued in *Betrayers of the Truth* by Broad and Wade [39], while allegations that the psychologist Cyril Burt acted fraudulently are examined carefully in [189].

The Invalid Theorem

Avoid the mistake of calling a theorem into question through sentences such as the following:

> The theorem holds for any continuously differentiable function f. Unfortunately, the theorem is invalid because S is not path connected.

A theorem holds and is valid, by definition. A theorem might be *applicable to any continuously differentiable function* or *its invocation* may be *invalid because S is not path connected.*

"This Paper Proves ... "

In the abstract and introduction it is tempting to use wording such as "this paper proves" or "Section 3 shows" in place of "we prove" or "we show". This usage grates on the ear of some readers, as it is logically incorrect (though "Theorem 2 gives" cannot be criticized). The grating can be avoided by rewriting, but care is required to avoid a succession of sentences beginning "we".

Chapter 7
Revising a Draft

Simply go through what you have written and
try to curb the length of sentences,
question every passive verb and if possible make it active,
prune redundant words, and look for nouns used instead of verbs.
— BERNARD DIXON, *Sciwrite* (1973)

Every single word that I publish I write at least six times.
— PAUL R. HALMOS, Interview in *Paul Halmos:*
Celebrating 50 Years of Mathematics (1991)

Two pages of the final manuscript . . .
Although they look like a first draft,
they had already been rewritten and retyped—like
almost every other page—four or five times.
With each rewrite I try to make what I have
written tighter, stronger and more precise,
eliminating every element that is not doing useful work.
Then I go over it once more, reading it aloud, and am always
amazed at how much clutter can still be cut.
— WILLIAM ZINSSER, *On Writing Well* (1990)

When one has finished a substantial paper there is commonly a
mood in which it seems that there is really nothing in it.
Do not worry, later on you will be thinking
"At least I could do something good then."
— J. E. LITTLEWOOD, *Littlewood's Miscellany*[14] (1986)

---

[14]See [34].

7.1. How to Revise

All writing benefits from revision. Your first attempt can always be made clearer, more concise, and more forceful. Effective revision is a skill that is acquired through practice. A prerequisite is an understanding of the principles of writing discussed in the previous chapters, together with the ability to spot when these principles have been violated and the skill and courage to rectify the weaknesses.

Put a draft aside for a few days, so that you can examine it with a fresh mind. Then analyse the draft in different ways. Read it aloud and listen to the rhythm. Read it at high speed. Does the text flow? Are the sentences of varying length? Read at the page level, focusing on the shape and the density of ink on the page (assuming the draft has been typeset). Is there a good balance between equations and text? Does the paper look inviting? Do the key ideas stand out? Knuth [164, p. 3] notes that "Many readers will skim over formulas on their first reading of your exposition. Therefore, your sentences should flow smoothly when all but the simplest formulas are replaced by 'blah' or some other grunting noise."

Most good writers work through several drafts—see the quotations at the beginning of the chapter. Some writers adopt the spiral or factorial technique of writing, whereby sections or chapters are written and revised in the order 1, 2, 1, 2, 3, 1, 2, 3, 4, ... Halmos [121] describes this as "the best way to start writing, perhaps the only way." But it is not always necessary to start at the beginning: for example, you can leave the introduction to last and begin by writing a numerical examples section or your conclusions. Every paper is different and versatility is the key.

Some writers have the ability to produce first drafts that need little revision. The amount of revision required depends partly on the stage in the research process at which the writing is begun. As mentioned in Chapter 1, writing can be an integral part of the research process.

I make no distinction between editing, revising and rewriting in my own work. I begin with the intention of making whatever changes are necessary to produce the best exposition I can. This may involve minor deletions, insertions and reorderings, or it may require complete rewrites of paragraphs and sections. As an extreme example, I deleted the original first paragraph of this chapter and started again from scratch. Although I made full use of computers in writing this book, I did most of my writing and revising at my desk by writing on a blank sheet or a printout of a draft; only occasionally did I compose at the keyboard. When marking corrections, I use a different colour for each pass through the document and my printouts usually finish up covered in coloured ink.

Write or print your drafts with wide margins and with a line spacing big enough for you to write between the lines. (In LaTeX, when the `article` style is used, a suitable line spacing can be achieved by putting the command `\def\baselinestretch{1.3}` before the `\begin{document}` command.) This will give you a lot of space to mark revisions, which in turn will encourage you to make them.

Figure 7.1 contains a check-list of questions to ask yourself when revising your writing.

Many of the points discussed above and in previous chapters are illustrated in the extracts in the rest of this chapter, most of which are taken from the mathematics and computer science literature. Minor changes have been made to protect the guilty.

7.2. Examples of Prose

▷ GraphX—An experimental graphics development environment research project

> *Comments.*
> This title has the dubious distinction of having five nouns in succession, the first four used adjectivally. It contains three abstract *-al* or *-ent* words. This is not a title to arouse interest.

▷ This type of response surface determination produces information describing not only the variable coordinates for the optimum but also indicates the nature of the response surface in the domain of interest.

> *Comments.*
> The "not only" and "also" clauses do not match. The simplest solution is to delete "indicates". A possible improvement is "The information produced by this type of response surface determination describes not only the variable coordinates for the optimum but also the nature of the response surface in the domain of interest." I would also want to reword the clumsy "response surface determination".

▷ Thus there arises the question of whether a central database can be created in a manner which is consistent with the need for efficiency and breadth of scope.

> *Comments.*
> (1) *Thus there arises the question* can be replaced by the shorter and more direct *The question arises*.
> (2) The which is a wicked one (see §4.14). In fact, the "which is" can be deleted.

▷ Can any words, phrases or sentences be deleted without loss of force or meaning (every such unit must earn its place)?

▷ Can you replace any long, woolly words with short, pithy ones (for example, require → need, utilize → use)?

▷ Are sentences and paragraphs in the best order, or should they be reordered or restructured?

▷ Are there any ambiguities or blatantly wrong words (such as *factor* instead of *feature*)?[a]

▷ Is there any unnecessary repetition?

▷ Can you convert a sentence from the passive to the active voice?

▷ Is every claim fully supported?

▷ Are the mathematical arguments and equations correct?

▷ Is the notation consistent? Can it be simplified or made more logical?

▷ Have quotations, references and numerical results been copied into the paper correctly?

▷ Is due reference made to the work of other authors (beware of "citation amnesia")?

▷ Are equations, results and the reference list properly numbered? Are all the cross-references and bibliographic citations correct?

---

[a]A draft of this book contained the phrase "chop sentences mercifully" instead of "chop sentences mercilessly"!

Figure 7.1. Check-list for revising.

▷ *Abstract*—This paper discusses the aims and methods of the FTEsol project and in this context discusses the architecture and design of the control system being produced as the focus of the project.

> *Comments.*
>
> (1) The phrases *in this context* and *as the focus of the project* serve no useful purpose and can be deleted.
>
> (2) As mentioned in §6.6, it is not a good idea to begin an abstract with *This paper.* A complete rewrite is needed: "The aims and methods of the FTEsol project are discussed, together with the architecture and design of the control system being produced." This version avoids the repeated *discusses* in the original. A more direct alternative, if the use of *we* is allowed, is "We discuss the aims and methods of the FTEsol project . . . ."

▷ `Nsolve` finds numerically the 5 complex solutions to this fifth-order equation.

> *Comments.*
>
> There is a lack of parallelism in *5* and *fifth*, which the reader may find disturbing. I would write "the five complex solutions to this fifth-order equation".

▷ Tables give a systematic and orderly arrangement of items of information. Tabular layout has the particular virtue of juxtaposing items in two dimensions for easy comparison and contrast. Tables eliminate tedious repetition of words, phrases and sentence patterns that can instead be put at the tops of columns and the sides of rows in the table. Although tables do not make much impact by visual display, it is possible, by careful arrangements, to emphasize and highlight particular items or groups of information.

> *Comments.*
>
> This is the first paragraph of a section on tables from a manual on technical writing. The first sentence reads like a dictionary definition and is surely not telling readers anything they do not know already. The ideas in the third and fourth sentences are not clearly expressed. Revised version:
>
> > Tables juxtapose items in two dimensions for easy comparison and contrast. Their row and column labels save the tedious repetition of words that would be necessary if the information were presented in textual form. Although tables lack the visual impact of

> graphs, information can be grouped and highlighted
> by careful use of rules and white space.

▷ In order to pinpoint the requirements for an effective microchip development environment sufficiently to definitively obtain answers to the above
questions, it is essential to be able to interview a wide variety of microchip
developers who are knowledgeable and experienced in such matters.

> *Comments.*
> (1) *In order* is superfluous.
> (2) *To definitively obtain* is a split infinitive (to-adverb-verb)
> that is more naturally written as *to obtain definitive answers*.
> It is probably better to replace the whole phrase by *to answer*.
> (3) *Essential to be able to* is probably better replaced by
> *necessary to*.

▷ Due to the advances in computer graphics and robotics, a new interest in
geometric investigations has now arisen within computer science focusing
mainly on the computational aspects of geometry, forming the research field
known as computational geometry.

> *Comments.*
> *A new interest . . . has now arisen* is a passive phrase that is
> easily improved. At the same time, we can remove the awkward
> and incorrect *due to* (see §4.14). The phrase *focusing . . . aspects
> of geometry* seems unnecessary. My revised version:
>
>> Advances in computer graphics and robotics have
>> stimulated a new interest in geometric investigation
>> within computer science, forming the research field
>> known as computational geometry.

▷ In terms of fractals, a straight line has a dimension of one, an irregular
line has a dimension of between one and two, and a line that is so convoluted
as to completely fill a plane has a dimension approaching the dimension of
the plane, namely a dimension of two. Fractal dimensions assign numbers
to the degree of convolution of planar curves.

> *Comments.*
> This extract is taken from a model paper given in a book
> about writing a scientific paper. The previous two sentences
> had also been about fractals. The phrase "in terms of fractals"
> is unnecessary and the order of these two sentences should be

reversed. "Fractal dimensions" do not "assign numbers" but *are* numbers. It is not clear, grammatically, whether the "namely a dimension of two" applies to the convoluted line or the plane. The split infinitive "to completely fill" can be replaced by "to fill". My rewritten version:

> Fractal dimensions are numbers that measure the degree of convolution of planar curves. A straight line has dimension one, an irregular line has a dimension between one and two, and a line that is so convoluted as to fill a plane has a dimension approaching two, which is the dimension of the plane.

▷ Data flow analysis determines the treatment of every parameter and COMMON variable by every subprogram with sufficient precision that non-portable parameter passing practices can be detected.

> *Comments.*
> This sentence is difficult to understand on its first reading. The phrase *by every subprogram* delays the punch-line and the *every*s also delay comprehension. It is not clear whether "with sufficient precision" applies to the subprograms or the analysis. In fact, the latter was intended. A better version is "Data flow analysis determines the treatment of parameters and COMMON variables with sufficient precision that nonportable parameter passing practices can be detected."

▷ [12] reports an eigenproblem from the automobile industry. The eigenvalues of interest are those ones having real part greater than zero.

> *Comments.*
> A citation makes a weak start to a sentence and jolts the eye. In the second sentence *ones* is unnecessary and the sentence can be made more concrete. Revised version:

> > Jones reports an eigenproblem from the automobile industry [12]. The eigenvalues of interest are those lying in the right half plane.

▷ Command names have been defined as two letter sequences because it is believed that users prefer to avoid verbosity.

Comments.

Who is the believer? *It is believed* is better replaced by *we believe*, or *evidence suggests*, or *our experience has shown*, preferably with a reference. A hyphen is needed in *two-letter sequences*, otherwise the meaning is *two sequences of letters*.

▷ When dealing with sets of simple figures, a basic problem is the determination of containment relations between elements of the set.

Comments.

When dealing is a dangling participle: we are not told who is dealing with the sets. A better version results on omitting *when dealing* and changing *the determination of* to *to determine*. Even better is

> A basic problem for sets of simple figures is to determine containment relations between elements of the set.

▷ In the function, $\exp(x)$ is first tested for overflow. If it does, then `inf` is returned.

Comments.

"If it does" refers to "tested for overflow" rather than "overflows", which the intended meaning requires. The second sentence can be replaced by "If overflow is detected then `inf` is returned."

▷ It is anticipated that the early versions of the system will provide definitive enough information that it will be reasonable to design with some assurance a variety of other systems which should be broader in scope.

Comments.

(1) *Anticipated* should be *expected*. This usage is described as "avoided by careful writers and speakers of English" in the *Collins English Dictionary* [60] but is so frequent that it may one day become accepted. Anticipate means to take action against (*The enemy had anticipated our move*).

(2) There are not degrees of definitiveness.

(3) *Reasonable* should probably be *possible*.

(4) There is a wicked which. The last phrase could be replaced by *systems of broader scope*.

▷ As far as the minimum eigenvalues of the other boundary element matrices are concerned they can only be small if the value of k is close to a corresponding element of S_Γ.

> *Comments.*
> "As far as ... is/are concerned" is a phrase more appropriate to speech than writing. This sentence can be shortened considerably: "The minimum eigenvalues of the other boundary element matrices can be small only if k is close to a corresponding element of S_Γ."

▷ This object is achieved by utilizing a set of properties which the signal is known or is hypothesized as possessing.

> *Comments.*
> This passive sentence can be shortened and made active: "We achieve this aim by using properties that we know or hypothesize the signal to possess." This simplified form shows that the sentence says little. The paragraph in which the sentence appears should be rewritten.

▷ The *main* purpose of any scientific article is to convey in the fewest *number of* words the ideas, procedures and conclusions of an investigator *to the scientific community*. Whether *or not* this *admirable* aim is accomplished depends *to a large extent* on how skillful the author is in *assembling the words of* the *English* language.

> *Comments.*
> These are the opening two sentences of a medical journal editorial titled "Use, Misuse and Abuse of Language in Scientific Writing". The italics are those of Gregory [117], who points out that the italicized words can be omitted without loss of meaning.

▷ Mathematica was found to be a suitable environment in which to perform the computational experiments.

> *Comments.*
> The following rewrite is much shorter, avoids the passive voice, and takes for granted the "suitability": "We carried out our computational experiments in Mathematica."

▷ These observations simply imply that nearby orbits separate from the orbit of γ after many iterations of the map G. Hardy and Littlewood (1979) prove a classical theorem that is useful in this context.

Comments.

There is nothing intrinsically wrong with this extract, but the astute writer may wish to rewrite to remove two minor infelicities. First, the near repetition in the phrase "simply imply" could distract the reader. Second, the eye naturally tends to read "iterations of the map G. Hardy"—symbols can cause trouble at the end of a sentence as well as at the beginning!

7.3. Examples Involving Equations

▷ *Theorem.* Let $A \in \mathbb{R}^{m \times n}$ be a given matrix, and let $A = U\Sigma V^T$ be a singular value decomposition of A. Then the problem

$$\max\{Re \ \mathrm{trace}(AQ): \ Q \in \mathbb{R}^{n \times n} \text{ is orthogonal}\}$$

has the solution $Q = VU^T$, and the value of the maximum is $\sigma_1(A) + \cdots + \sigma_n(A)$, where $\{\sigma_i(A)\}$ is the set of singular values of A.

Comments.

The statement of this theorem is clear and readable but it can be improved. The first sentence can be shortened to "Let $A \in \mathbb{R}^{m \times n}$ have the singular value decomposition $A = U\Sigma V^T$." The second sentence can be rewritten as "Then

$$\max\{Re \ \mathrm{trace}(AQ): \ Q \in \mathbb{R}^{n \times n} \text{ is orthogonal}\} = \sum_{i=1}^{n} \sigma_i,$$

where the σ_i are the singular values of A, and the maximum is attained when $Q = VU^T$." This is shorter and, arguably, easier to read. Even better is to define the σ_i at the beginning by appending "where $\Sigma = \mathrm{diag}(\sigma_i)$" to the first sentence.

▷ Writing $\theta_i = (E|x| + f)_i$ we have

$$\rho(x) \leq \| \, |A^{-1}| \, \mathrm{diag}(\theta_i) \, \|_\infty.$$

Using properties of the ∞-norm we obtain

$$\rho(x) \leq \| \, |A^{-1}|(E|x| + f) \, \|_\infty.$$

Comments.

The presentation hides the fact that the upper bound in the second inequality is equal to that in the first. The following rewrite is more informative:

Writing $\theta_i = (E|x| + f)_i$, and using properties of the ∞-norm, we obtain

$$\rho(x) \leq \| \, |A^{-1}| \operatorname{diag}(\theta_i) \, \|_\infty$$
$$= \| \, |A^{-1}|(E|x| + f) \, \|_\infty.$$

▷ Let A be $n \times n$. Show that if for any Hermitian matrix H, $\operatorname{trace}(HA) = 0$, then $A = 0$.

Comments.
This question is ambiguous because *any* can mean *whichever* or *at least one* in everyday English. As Halmos [121] recommends, *any* should always be replaced by *each* or *every* in mathematical writing.

▷ Suppose now that the assumption $a = 1$ fails.

Comments.
An assumption does not fail or succeed: it is either invalid or valid. Better wording might be *Suppose now that the condition $a = 1$ is not satisfied*, or *Suppose now that $a \neq 1$*.

▷ **Introduction**
Throughout this paper $\|A\|_F$ denotes the Frobenius norm $\left(\sum_{i,j} a_{ij}^2\right)^{1/2}$ of a real-valued matrix A and A^+ denotes the Moore–Penrose pseudo-inverse of A. We define the set

$$\mathcal{U}_+ = \{\, U \in \mathbb{R}^{n \times n} : U^T U = I, \; \det(U) = 1 \,\}.$$

Given $A, B \in \mathbb{R}^{m \times n}$ we are interested in the *orthogonal Procrustes problem* in its pure rotation form:

$$\min_{U \in \mathcal{U}_+} \|A - BU\|_F.$$

Comments.
This is a weak start. It can be improved by stating the problem in the first sentence. We can delay the definition of A^+ until it is first used. "Real-valued" can be shortened to "real", but we can dispense with it altogether by using the $\mathbb{R}^{m \times n}$ notation. The "Given" phrase does not read well—interest in the mathematical problem is surely independent of being given matrices A and B. Revised first sentence:

The pure rotation form of the *orthogonal Procrustes problem* is

$$\min_{U \in \mathcal{U}_+} \|A - BU\|_F, \qquad A, B \in \mathbb{R}^{m \times n},$$

where

$$\mathcal{U}_+ = \{\, U \in \mathbb{R}^{n \times n} : U^T U = I, \ \det(U) = 1 \,\}$$

and $\|A\|_F = \left(\sum_{i,j} a_{ij}^2 \right)^{1/2}$ is the Frobenius norm.

The writer should now go on to mention applications in which this problem arises, explain what is known about its solution, and state the purpose of the paper.

▷ *Theorem.* Let A be an $n \times p$ complex matrix with rank p. We define the $p \times p$ positive definite Hermitian matrix $S = A^*A$ and the $n \times p$ matrix $\overline{Q} = AS^{-\frac{1}{2}}$. Let $\mathcal{U}$ be the set of all $n \times p$ orthonormal matrices. Then the following is true:

$$\overline{Q} \in \mathcal{U} \qquad \text{and} \qquad \|A - \overline{Q}\|_2 \le \min_{Q \in \mathcal{U}} \|A - Q\|_2.$$

Comments.

(1) The rank assumption implies $p \le n$. It is clearer, though less concise, to say "Let A be an $n \times p$ complex matrix with $n \ge p = \operatorname{rank}(A)$."

(2) There is no need to introduce the symbol S. If the existence of $(A^*A)^{-1/2}$ is thought not to be obvious (note the preferable slashed fraction in the exponent), it can be established in the proof. The symbol $\mathcal{U}$ can also be dispensed with.

(3) The inequality is an equality. The second sentence onwards can be simplified as follows: Let $\overline{Q} = A(A^*A)^{-1/2}$. Then $\overline{Q}$ is orthonormal and

$$\|A - \overline{Q}\|_2 = \min\{\, \|A - Q\|_2 : Q^T Q = I_p \,\}.$$

▷ The optimality of the constant $\pi \log n / 4$ in inequality (14.3) is due to Smith [10].

Comments.

This sentence suggests that Smith made the constant optimal. Better is *was first established by Smith.*

▷ (From an abstract:) The bound is derived in the case of k $(0 \leq k \leq p)$ explanatory variables measured with error.

> *Comments.*
> The intrusive inequalities and the all-purpose phrase "in the case of" can be removed, and the reader told, or reminded, what p is, by writing "The derivation of the bound allows for any k of the p explanatory variables to be measured with error."

7.4. Examples from My Writing

Here are some examples from my own writing of how I improved drafts.

(1) Original first two sentences of paper:

▷ Summation of floating point numbers is a ubiquitous and fundamental operation in scientific computing. It is required when evaluating inner products, means, variances, norms, and all kinds of functions in nonlinear problems.

> Improved version (shorter, less passive, more direct):
>
> > Sums of floating point numbers are ubiquitous in scientific computing. They occur when evaluating inner products, means, variances, norms, and all kinds of nonlinear functions.
>
> An alternative (avoids the dangling participle "when evaluating", at the cost of a more passive construction):
>
> > Sums of floating point numbers occur everywhere in scientific computing: in the evaluation of inner products, means, variances, norms, and all kinds of nonlinear functions.

(2) Original:

▷ Here, instead of immediately feeding each correction e_i back into the summation, the corrections are accumulated as $e = \sum_{i=1}^{n} e_i$ (by recursive summation) and then the global correction e is added to the computed sum.

> Improved version (omits the unnecessary mathematical notation):

Here, instead of immediately feeding each correction back into the summation, the corrections are accumulated by recursive summation and then the global correction is added to the computed sum.

(3) Original first sentence of abstract:

▷ If a stationary iterative method is used to solve a linear system $Ax = b$ in floating point arithmetic how small can the method make the error and the residual?

Improved version (avoids the misleading *if*, more direct):

How small can a stationary iterative method for solving a linear system $Ax = b$ make the error and the residual in the presence of rounding errors?

(4) Original first sentence of paper:

▷ A *block algorithm* in matrix computations is one that is defined in terms of operations on submatrices rather than matrix elements.

A copy editor removed the words *one that is*. This changes the meaning, since the sentence now states a property rather than gives a definition, but I felt that the shorter sentence was an improvement.

7.5. A Revised Proof

Gershgorin's theorem, a well-known theorem in numerical linear algebra, specifies regions in the complex plane in which the eigenvalues of a matrix must lie. Here is the theorem and a proof that is correct, but can be improved.

Theorem 1 (Gershgorin, 1931) *The eigenvalues of $A \in \mathbb{C}^{n \times n}$ lie in the union of the n disks in the complex plane*

$$D_i = \{ z \in \mathbb{C} : |z - a_{ii}| \leq \sum_{\substack{j=1 \\ j \neq i}}^{n} |a_{ij}| \}, \qquad i = 1, \ldots, n.$$

Proof. The proof is by contradiction. Let λ be an eigenvalue of A and x an eigenvector associated with λ and assume that $\lambda \notin D_i$ for $i = 1, \ldots, n$. Then from

$$Ax = \lambda x$$

and

$$(\lambda I - A)x = 0$$

we have

$$(\lambda - a_{ii})x_i = \sum_{\substack{j=1 \\ j \neq i}}^{n} a_{ij}x_j, \qquad i = 1, \ldots, n.$$

Taking absolute values and applying the triangle inequality gives

$$|\lambda - a_{ii}||x_i| \leq \sum_{\substack{j=1 \\ j \neq i}}^{n} |a_{ij}||x_j|, \qquad i = 1, \ldots, n.$$

Assume that $|x_k| = \max_i |x_i|$. Then, dividing the kth inequality by $|x_k|$, we have

$$|\lambda - a_{kk}| \leq \sum_{\substack{j=1 \\ j \neq k}}^{n} |a_{kj}||x_j|/|x_k| \leq \sum_{\substack{j=1 \\ j \neq k}}^{n} |a_{kj}|,$$

showing that λ is contained in the disk $\{\lambda : |\lambda - a_{kk}| \leq \sum_{j \neq k}^{n} |a_{kj}|\}$, which is a contradiction. $\square$

Several different proofs of Gershgorin's theorem exist, and this one is the most elementary and direct. The presentation of the proof has several failings, though.

1. The proof is too detailed and contains some unnecessary equations and inequalities.

2. The proof by contradiction is unnecessary, since the assumption that is to be contradicted is not used in the proof.

3. The "assumption" on k is really a definition of k and is clearer if phrased as such.

4. In the last line of the proof the disk can be described by its name, D_k.

The following proof uses the same reasoning but is much more concise and no less clear.

Proof. Let λ be an eigenvector of A and x a corresponding eigenvector, and let $|x_k| = \max_i |x_i|$. From the kth equation in $Ax = \lambda x$ we have

$$\sum_{\substack{j=1 \\ j \neq k}}^{n} a_{kj} x_j = (\lambda - a_{kk}) x_k.$$

Hence

$$|\lambda - a_{kk}| \leq \sum_{\substack{j=1 \\ j \neq k}}^{n} |a_{kj}||x_j|/|x_k|,$$

and since $|x_j|/|x_k| \leq 1$ it follows that λ belongs to the kth disk, D_k. $\square$

Of course a proof has to be written with the intended audience in mind. For an undergraduate text the revised proof is probably too concise and some intermediate steps could be added.

7.6. A Draft Article for Improvement

Below is a shortened version of an article that I wrote for an undergraduate mathematics magazine. I have introduced over twenty errors of various kinds, though most are relatively minor. How many can you spot? If you are an inexperienced writer, criticizing this "draft" will be a valuable exercise.

Numerical Linear Algebra in the Sky

In aerospace computations, transformations between different co-ordinate systems are accomplished using the *direction cosine matrix* (DCM), which is defined as the solution to a time dependent matrix differential equation. The DCM is 3×3 and exactly orthogonal, but errors in computing it lead to a loss of orthogonality. A simple remedy, first suggested in a research paper in 1969 is to replace the computed DCM by the nearest orthogonal matrix every few steps. These computations are done in real-time by an aircrafts on-board computer so it is important that the amount of computation be kept to a minimum. One suitable method for computing a nearest orthogonal matrix is described in this Article. We begin with the case of 1×1 matrices—scalars.

(a) Let x_1 be a nonzero real number and define the sequence:

$$x_{k+1} = \frac{1}{2}(x_k + 1/x_k), \qquad k = 1, 2, \ldots, .$$

If you compute the first few terms on your calculator for different x_1 (e.g. $x_1 = 5$, $x_1 = -3$) you'll find that x_k converges to ± 1; the sign depending on the sign of x_1. Prove that this will always be the case (*Hint:* relate $x_{k+1} \pm 1$ to $x_k \pm 1$ and then divide this two relations). This result can be interpreted as saying that the iteration computes the nearest real number of modulus one to x_1.

(b) This scalar iteration can be generalized to matrices without loosing it's best approximation property. For a given nonsingular $X_1 \in \mathbb{R}^{n \times n}$ define

$$X_{k+1} = \frac{1}{2}(X_k + X_k^{-T}) \tag{1}$$

(This is one of those very rare situations where it really is necessary to compute a matrix inverse!) Here, X_k^{-T} denotes the transpose of the inverse of X_k. Natural questions to ask are: Is the iteration well defined (i.e., is X_k always nonsingular)? Does it converge? If so, to what matrix?

To investigate the last question suppose that $X_k \rightarrow X$. Then X will satisfy $X = \frac{X+X^{-T}}{2}$, or $X = X^{-T}$, or

$$X^T X = I,$$

thus X is orthogonal! Moreover, X is not just any orthogonal matrix. It is the nearest one to X_1 as shown by the following

Theorem 1

$$\|X - X_1\| = \min\{\|Q - X_1\| : Q \in R^{n \times n}, \; Q^T Q = I_n\}$$

where the norm is denoted by

$$\|X\| = \left(\sum_{i}^{n} \sum_{j=1}^{n} x_{ij}^2 \right)^{1/2}.$$

This is the matrix analogue of the property stated in (a).

Returning to the aerospace application, the attractive feature of iteration (1) is that if we don't wait to long before "re-orthogonalising" our computed iterates then just one or two applications of the iteration (1) will yield the desired accuracy with relatively little work. $\square$

Here are the corrections I would make to the article. (In repeating the exercise myself some time after preparing this section, I could not find all the errors!)

1. First paragraph: hyphenate *time-dependent*; comma after 1969; *aircraft's*; comma after *on-board computer*; *article* in lower case.

2. Second paragraph: no colon after *sequence*. In display, $\frac{1}{x_k}$ instead of $1/x_k$, and replace ".....," by "....". Comma after *e.g.* and instead of semicolon after ± 1. *These two relations*; *modulus* 1.

3. Third paragraph: *losing*; *its*. Right parenthesis in display (1) is too large and full stop needed at end of display.

4. Fourth paragraph: third X should be in mathematics font, not roman; $(X + X^{-T})/2$. The equation $X^T X = I$ should not be displayed and it should be followed by a semicolon instead of a comma. (The spacing in $X^T X$ should be tightened up—see page 192 for how to do this in LATEX.) Comma after X_1; *following theorem*.

5. No need to number the theorem as it's the only one. It should begin with words: *The matrix X_1 satisfies*. $\mathbb{R}^{n \times n}$ instead of R^{nxn} (two changes). Comma at end of first display. "I_n" is inconsistent with "I" earlier: make both I. *Denoted* should be *defined*. In first sum of second display, $i = 1$. The parentheses are too large.

6. Last paragraph: *too long*. Wrong opening quotes. For consistency with *generalized* (earlier in article), spell as *re-orthogonalizing*. Logical omission: I haven't shown that the iteration converges, or given or referred to a proof of the theorem.

Chapter 8
Publishing a Paper

In the old days, when table making was a handcraft,
some table makers felt that every entry in a table was a theorem
(and so it is) and must be correct. . . .
One famous table maker used to put in errors deliberately
so that he would be able to spot his work
when others reproduced it without his permission.

— PHILIP J. DAVIS, *Fidelity in Mathematical Discourse:*
Is One and One Really Two? (1972)

The copy editor is a diamond cutter who refines and polishes,
removes the flaws and shapes the stone into a gem.
The editor searches for errors and inaccuracies, and prunes the useless,
the unnecessary qualifiers and the redundancies.
The editor adds movement to the story by substituting active
verbs for passive ones, specifics for generalities.

— FLOYD K. BASKETTE, JACK Z. SISSORS and
BRIAN S. BROOKS, *The Art of Editing* (1992)

Lotka's law states that the number of people producing n papers
is proportional to $1/n^2$.

— FRANK T. MANHEIM, *The Scientific Referee* (1975)

Memo from a Chinese Economic Journal:
We have read your manuscript with boundless delight.
If we were to publish your paper,
it would be impossible for us to publish any work of lower standard.
And as it is unthinkable that in the next thousand years
we shall see its equal, we are, to our regret,
compelled to return your divine composition,
and to beg you a thousand times to overlook our short sight and timidity.

— From *Rotten Rejections* (1990)

Once your paper has been written, how do you go about publishing it? In this chapter I describe the mechanics of the publication process, from the task of deciding where to submit the manuscript to the final stage of checking the proofs. I do not discuss how to decide whether your work is worth trying to publish, but Halmos offers some suggestions (particularly concerning what not to publish) in [124]. When to publish is an important question on which it is difficult to give general advice. I recommend that you find out the history of some published papers. Authors are usually happy to explain the background to a paper. *Current Contents* (see §14.3) regularly carries articles describing the background to a "Citation Classic", which is a paper that has been heavily cited in the literature. A good example is the article by Buzbee [47] describing the story of the paper "On direct methods for solving Poisson's equations" [B. L. Buzbee, G. H. Golub, and C. W. Nielson. *SIAM J. Numer. Anal.*, 7(4):627–656, 1970]. The article concludes with the following comments.

> So, over a period of about 18 months, with no small amount of mathematical sleuthhounding, we completed this now-*Classic* paper. During that 18 months, we were tempted on several occasions to publish intermediate results. However, we continued to hold out for a full understanding, and, in the end, we were especially pleased that we waited until we had a comprehensive report.

I concentrate here on publishing in a refereed journal. Another important vehicle for publication is conference proceedings. These are more common in computer science than mathematics, and in computer science some conference proceedings are at least as prestigious as journals. It is important to realize that many conference proceedings and a few journals are not refereed, and that when you are considered for hiring or promotion, refereed publications will probably carry greater weight than unrefereed ones.

8.1. Choosing a Journal

There are more journals than ever to which you can send a scientific paper for publication, so how do you choose among them? The most important question to consider is which journals are appropriate given the content of the paper. This can be determined by looking at recent issues and reading the stated objectives of the journal, which are often printed in each issue. Look, too, at your reference list—any journals that are well represented are candidates for submission of your manuscript. Experts in your area will also be able to advise on a suitable journal.

Several other factors should be considered. One is the prestige and quality of the journal. These rather hard-to-judge attributes depend chiefly on the standard of the papers published, which in turn depends on the standard of the submissions. The higher quality journals tend to have lower acceptance rates, so publishing in these journals is more difficult. Acceptance rates are usually not published, but may be known to members of editorial boards. The figures sent by SIAM to its editors do not give acceptance rates, but they suggest that the rates for SIAM journals are usually between 30% and 50%, depending on the journal.

Gleser [106] states that "the major statistical journals receive many more manuscripts than they can eventually publish and, consequently, have a high rate of rejection", and he remarks that *The Journal of the American Statistical Association* rejects nearly 80% of all papers submitted.

One way to quantify the prestige and quality of a journal is to look at how often papers in that journal are cited in the literature [89], [90]. Such information is provided by the *Journal Citation Reports* published by the Institute for Scientific Information. A study of mathematics journals based on the 1980 report of citation statistics is given by Garfield [92]; his article identifies the fifty most-cited mathematics journals, the most-cited papers from the most-cited journals, and the journals with the highest impact factor (a measure of how often an average article is cited [100]). Based on the 1980 data, the journals with the ten highest impact factors are, from highest to lowest, *Comm. Pure Appl. Math., Ann. Math., Adv. in Math., SIAM Review, Acta Math.—Djursholm, Invent. Math., SIAM J. Numer. Anal., Stud. Appl. Math., Duke Math. J., Math. Program.*

The circulation of a journal should also be considered. If you publish in a journal with a small circulation your paper may not be as widely read as you would like. Relatively new journals from less-established publishers are likely to have small circulations, especially in the light of the budget restrictions imposed on many university libraries. SIAM publishes circulation information in the first issue of the year of each journal; see Table 8.1. The circulation of *SIAM Review* is so high because every SIAM member receives it. For journals published electronically and not requiring a subscription, circulation may be of less concern.

The audience for your paper depends very much on the journal you choose. For example, a paper about numerical solution of large, sparse eigenvalue problems could be published in the *SIAM Journal on Scientific Computing*, where it would be seen by a broad range of workers in scientific computing; in the *SIAM Journal on Matrix Analysis and Applications*, whose readership is more biased towards pure and applied linear algebra; or in the *Journal of Computational Physics*, whose readership is mainly physicists and applied mathematicians, many of whom need to solve practical

Table 8.1. Circulation figures for some SIAM journals.

Journal	Total Distribution Per Issue, 1997	Issues Per Year
SIAM J. Appl. Math.	2485	6
SIAM J. Comput.	2069	6
SIAM J. Control Optim.	1965	6
SIAM J. Math. Anal.	1612	6
SIAM J. Matrix Anal. Appl.	1659	4
SIAM J. Numer. Anal.	2453	6
SIAM J. Optim.	1450	4
SIAM Review	11531	4
SIAM J. Sci. Comput.	2150	6

eigenvalue problems.

Other factors to consider when choosing a journal are the delays, first in refereeing. How long you have to wait to receive referee reports varies among journals. It depends on how much time the journal allows referees, how efficient an editor is at prompting tardy referees, and, of course, it depends on the referees themselves (who usually act as referees for more than one journal).

The other major delay is the delay in publication: the time from when a paper is accepted to when it appears in print. For a particular article, this delay can be calculated by comparing the date of final submission or acceptance (displayed for each article by most journals) with the cover date of the journal issue. The publication delay depends on the popularity of the journal and the number of pages it publishes each year. A survey of publication delays for mathematics journals appears each year in the *Notices of the American Mathematical Society* journal ("Backlog of Mathematics Research Journals")—it makes interesting reading.

The publication delay also depends, partly, on the author. Ervin Rodin, the editor-in-chief of *Computers and Mathematics with Applications*, explains in an editorial [238] some of the reasons for delays. Four that are not specific to this particular journal are that the figures or graphs are not of high enough quality, the references are not given in full detail, the equations are inconsistently numbered, and the proofs are not returned promptly.

Finally, if your paper has been prepared in TeX (see Chapter 13) you might prefer to send it to a journal that typesets directly from author-supplied TeX source; as well as saving on proofreading this sometimes brings the benefit of extra free reprints (nowadays most journals provide some free reprints for all papers—as few as 15 or as many as 100).

8.2. Submitting a Manuscript

Before submitting your manuscript (strictly, it is a paper only after it has been accepted for publication) you should read carefully the instructions for authors that are printed in each issue of your chosen journal. Most of the requirements are common sense and are similar for each journal. Take particular note of the following points.

1. To whom should the manuscript be submitted? Usually it should be sent to the editor-in-chief, but some journals allow manuscripts to be sent directly to members of the editorial board; judgement is required in deciding whether to take advantage of this option (it may be quicker, since the manuscript skips the stage where an editor-in-chief selects an associate editor). Usually, the editorial addresses are printed in each issue of a journal. Look at a recent issue, as editors and their addresses can change.

2. Enclose a covering letter that states to which journal the manuscript is being submitted—this is not obvious if the organization in question has several journals, as does SIAM. State the address for correspondence if there is more than one author. Usually only the designated author receives correspondence, proofs and reprint order forms. If your address will change in the foreseeable future, say so, even if the change is only temporary. (This is particularly important at the typesetting stage, after a paper has been accepted, because proofs must be dealt with quickly.)

3. How many copies of the manuscript are required? SIAM requires five. If the destination is abroad, send them by air mail; don't use surface mail, which can take several weeks. Even if a paper is rejected, the manuscript is not usually returned. You should keep a copy, particularly as you may need it at the proofreading stage.

4. Always submit single-sided copies (not double-sided) and fasten them with a staple to avoid pages being lost (the referees can easily remove the staple if they wish). Provide your full address on the manuscript (some authors forget). Dating the manuscript may help to prick the conscience of a tardy editor or referee.

5. Give key words and the Mathematics Subject Classifications and Computing Reviews classification, if these are required by the journal.

It is a good policy to include them as a matter of course. You may also include at this stage a "running head"—a shortened title that appears at the tops of pages in the published paper. The running head should not exceed about 50 characters.

6. If your manuscript cites any of your unpublished work it will help the referees if you enclose a copy of that work, particularly if the manuscript relies heavily on it. Doing so avoids delays that might result from the referees asking to see the unpublished work.

7. If you are using LaTeX double check that the cross-references and citations are correct. Adding a new equation or reference at a late stage and not running LaTeX twice (or three times if you are using BibTeX—see Table 13.2) can result in incorrect numbering. I have seen citation errors of this type persist in a published journal article. Also, if the journal accepts TeX papers, use the style files or macros provided by the publisher; you may still be able to convert the paper to the required format once it has been accepted.

8. Most journals require that any material submitted for review and publication has not been published elsewhere. Papers that have appeared in preliminary form in conference proceedings are usually an exception to this rule. SIAM, for instance, requires that papers that have appeared in conference proceedings or in print anywhere in an abbreviated form be significantly revised before they are submitted to a SIAM journal. If your paper has already appeared in published form you must make this clear when the paper is submitted, and you must indicate so by a footnote on the first page.

9. Before putting your manuscripts in the envelope, check that no pages are missing from any of the copies. Again, delays will result if one of the copies is incomplete.

10. You should receive an acknowledgement of receipt of the manuscript within four weeks of submitting it. If you do not, write to ask whether the manuscript was received.

8.3. The Refereeing Process

I will explain how the refereeing process works for SIAM journals (this discussion is partly based on an article by Gear [103]). Procedures for most other journals are similar. If a manuscript is submitted to "The Editor" at the SIAM office, it is logged by SIAM and a letter of acknowledgement is sent to the author, giving a manuscript number that should be quoted in future correspondence. The manuscript is then passed on to the editor-in-chief, who assigns it to a member of the editorial board (this stage can take a few weeks). SIAM (or in some cases, the editor-in-chief) mails the submission, the covering letter, and a Manuscript Transmittal Sheet to the chosen editor. The editor writes to two or more people asking them to referee the paper, suggesting a deadline about six weeks from the time they receive the paper.

The editor may send a sheet of "instructions for referees". Figure 8.1 contains an extract from the SIAM instructions (many of the SIAM journals also have more specific instructions), which indicates again the points to consider before submitting a manuscript.

When all the referee reports have been received the editor decides the fate of the paper, informs the author, and notifies SIAM. (In some journals, including some of those published by SIAM and The Institute of Mathematics and Its Applications, the editor makes a recommendation to the editor-in-chief, who then writes to the author, possibly not naming the editor.) The paper can be accepted, accepted subject to changes, returned to the author for a substantial revision, or rejected.

If a referee report is not received on time the editor reminds the referee that the report is due. After six months of inactivity the manuscript is classed as "flagged" by SIAM, and the editor is urged by SIAM to expedite the current stage of the refereeing process. If you have had no response after six months you are quite entitled to contact SIAM or the editor-in-chief to enquire whether any progress has been made in refereeing the paper. Papers are sometimes mislaid or lost and your enquiry will at least reveal whether this has happened.

Few papers are accepted in their original state, if only because of minor typographical errors (typos). When preparing a revised manuscript in response to referee reports it is important to address all the points raised by the referees. In the covering letter for the revised version these points might be summarized, with an indication for each one of the action (if any) that was taken for the revision. If you do not act on some of the referees' recommendations you need to explain why. It greatly helps the editor if you explain which parts of the paper have been changed, as it is an irksome task to compare two versions of a paper to see how they differ. When you submit a revised manuscript it is a good idea to mark on the front page "Revised manuscript for journal X" together with the date. This will prevent the revised version from being confused with the original. When a revised paper is received, the editor may ask the referees to look at it or may make a decision without consulting them.

Keep in mind that the editor and referees are usually on your side. They

The most important criterion for acceptance of a publication is originality and correctness. Clear exposition and consistent notation are also required. All papers should open with an introduction to orient the reader and explain the purpose of the paper.

You are asked to prepare an unsigned report, in duplicate. We ask that you keep the report formal and impersonal so that the editor can forward it to the author.

A specific recommendation for acceptance or rejection should be excluded from the report. The following checkpoints are suggested for explicit consideration:

- Is the paper consistent with editorial objectives?

- Is the work correct and original or of wide appeal, if a review paper?

- Is its presentation clear and well organized?

- Is the notation well conceived and consistent?

- How does the paper relate to current literature?

- Are the references complete, relevant, and accurate?

- Does the title accurately characterize the paper?

- Does the abstract properly summarize the paper without being too vague?

- Does the introduction relate the paper to contemporary work and explain the purpose of the paper?

- Are equation numbers and figure numbers consistent?

When the manuscript fails to meet some explicit requirement, what material should the author develop to improve the presentation?

Cover Letter

Please return your report with a cover letter stating your recommendation concerning disposition of the paper. We ask that you justify a recommendation of acceptance as well as one of rejection, and please send the cover letter, report, and manuscript to the editor who requested this review.

Figure 8.1. Extract from SIAM instructions for referees.

are mostly busy people and would like nothing more than to be able to read your paper, quickly realize that it is correct and deserves publishing, and make that recommendation or decision. Anything you can do to help them is to your benefit. A major advantage of writing in a clear, concise fashion is that your papers may be refereed and edited more quickly!

8.4. How to Referee

The main task of a referee is to help an editor to decide whether a paper is suitable (or will be suitable, after revision) for publication in a journal. Opinions vary on precisely what a referee should do (see the references below), and different referees go about the task in different ways. It is useful to think of the refereeing process as comprising two stages, even though these are often combined. The first stage is an initial scrutiny in which the referee forms an overall view of the paper, without reading it in detail. The question to be considered is whether the paper is original enough and of sufficient interest to the journal's readers to merit publication *assuming* that the paper is free of errors. If a negative conclusion is reached, then there is no need for the referee to check the mathematical details.

If the overview reveals a paper potentially worth publishing then more detailed study is required, including consideration of the questions listed in Figure 8.1. How carefully the referee should check the mathematics depends partly on the nature of the paper. A paper applying a standard technique to a new problem may need less meticulous checking than one developing a new method of analysis. The referee's time is better concentrated on looking at the key ideas and steps in proofs rather than the low-level details, as this is the way that errors are most likely to be spotted.

The referee has to examine all facets of a paper and decide which lines of investigation will be the most fruitful in coming to a decision about the paper's merit. Experienced referees learn a lot from small clues. If one lemma is imprecisely stated or proved, perhaps other results need to be carefully checked. If an important relevant paper is not cited, perhaps the author is not fully conversant with the existing literature. An unreasonable assumption, perhaps hidden in a piece of analysis, calls into question the value of a result and may lead a referee to an immediate recommendation of rejection.

Some particular advice on refereeing follows.

1. If you are not willing or able to provide a report by the date requested by the editor, return the manuscript immediately. Alternatively, give the editor a date by which you can guarantee a report and ask if that would be acceptable.

2. You can save a lot of time by taking a global view of a paper before starting to examine the details. Twenty minutes spent coming to the suspicion that a paper appears flawed may enable you to pinpoint the errors much sooner than if you read the paper from start to finish, checking every line.

3. Make your recommendation in a cover letter to the editor, but not in the report itself. The report should contain the reasoning that supports your recommendation.

4. It is not necessary for you to summarize the paper if your summary would be similar to the abstract. If you can give a different and perhaps more perceptive overview of the work it will be of much use to the editor.

5. In considering changes to suggest for improving a paper that you think merits publication, you can ask many of the same questions that you would ask when writing and revising your own work. In particular, consider whether the notation, organization, length and bibliography are suitable.

6. If your recommendation is rejection, do not spend much time listing minor errors, typographical or otherwise, in the report (unless
   you wish, for example, to help an author whose first language is not
   English improve his or her grammar and spelling). Concentrate on
   describing the major flaws. Always try to offer some positive comments, though—imagine how you would feel on reading the report if
   the paper were yours.

7. Always be polite and avoid the use of language that can be interpreted
   as offensive or over-critical. In particular, avoid using unnecessary
   adjectives. Thus say "incorrect" rather than "totally incorrect" and
   "unwarranted" rather than "completely unwarranted".

8. Building a reputation as an efficient, conscientious and perceptive
   referee is worthwhile for several reasons. You will probably receive
   important papers to referee, thereby finding out about the work before most other researchers. As a trusted referee, your recommendations will help to influence the direction of a journal. You may
   even be invited onto editorial boards because of your reputation as a
   referee. Reputations are not made, though, by providing shallow reports of the form "well written paper, interesting results, recommend
   publication" when other referees point out serious weaknesses and
   recommend rejection. As Lindley [179] puts it, "A sound dismissal is
   harder to write than an advocacy of support."

For further discussion of the refereeing process I recommend the papers by Gleser [106], Lindley [179], Manheim [193], Parberry [216], Smith
[250] (intended for "applied areas of computer science and engineering")
and Thompson [272]. See also [164, §§15–17] and the collection *Publish
or Perish* [142], which includes chapters titled "The Refereeing Process",
"The Editor's Viewpoint" and "The Publisher's Viewpoint".

8.5. The Role of the Copy Editor

After a manuscript is accepted for publication it goes to a copy editor. The
copy editor of journal papers has three main aims: to do limited rewriting
or reorganizing of material in order to make the paper clear and readable;
to edit for correctness of grammar, syntax and consistency; and to impose
the house style of the journal (a fairly mechanical process). The copy editor
tries to make only essential changes and to preserve the author's style.

When you see a copy-marked manuscript for the first time your reaction might be one of horror at the mass of pencilled changes. Most of these

will be instructions to the printer to set the paper in the house style: instructions on the fonts to use, spacing in equations, placement of section headings, and a host of other details. There may also be some changes to your wording and grammar. The editor will have had a reason for making each change. The editor will probably have made improvements that you overlooked, such as finding words that are unnecessary, improving the punctuation, and correcting inconsistencies (copy editors keep track of words or phrases that have odd spellings, capitalization or hyphenation and make sure that you use them consistently). If you are unhappy with any of the changes a copy editor has made you can attach an explanatory note to the proof, or, if you are sure an error was made, reverse the change on the proof. Copy editors are always willing to reconsider changes and will pay attention to the author's views in cases of disagreement.

8.6. Checking the Proofs

Some time after your manuscript has been accepted, and after you have received and signed a copyright transferral form, you will receive page or galley proofs (galleys are sheets that have not been broken up into pages—if the journal typesets in TeX the galley stage may be nonexistent). You are asked to check these and return the marked proofs within a short period, often two days. For some journals, the original copy-edited manuscript is enclosed, so that you can see which changes the copy editor marked. If the marked manuscript is not enclosed you are at a disadvantage and you should check against your own copy of the manuscript, particularly for omissions, which can be very difficult to spot when you read only the proofs. The proofs you receive are usually photocopies, so if you see imperfections such as blotches and faint characters these might not be present on the original copy.

A thorough check of the proofs requires one read-through in which you do a line by line comparison with the marked manuscript, and another in which you read the proofs by themselves "for meaning". In addition to checking line by line, do a more global check of the proofs: look at equation numbers to make sure they are in sequence with no omissions, check for consistency of the copy editing and typesetting (typefaces, spacing, etc.) and make sure the running heads are correct. As mentioned above, most journals print the date the manuscript was first received and the date it was accepted. Check these dates, as they may help to establish priority if similar work is published by other researchers.

Some specific errors to check for are shown in Figure 8.2. If your paper contains program listings they should be checked with extra care, as printers find them particularly difficult to typeset. Common errors in typeset

- Unmatched parentheses: $a = (b + c/2$.

- Wrong font: $a = (\mathbf{b} + c)/2$.

- Missing words or phrases.

- Misspelt or wrong words (e.g., complier for compiler, special for spectral).

- Repeated words.

- Missing punctuation symbols (particularly commas).

- Incorrect hyphenation. If your manuscript contains a word hyphenated at the end of a line, the copy editor and printer may not know whether the hyphen is a permanent one or a temporary one induced by the line break.

- Widow: short last line of a paragraph appearing at the top of a page. Or "widow word": last word of a paragraph appearing on a line by itself (can be cured in TₑX by binding the last two words together with a tie).

- O (capital Oh) for 0 (zero), l (lower case ell) for 1, wrong kind of asterisk (* instead of $*$).

- Bad line breaks in mathematical equations.

- Incorrectly formatted displayed equations (e.g., poor alignment in a multiline display).

- Change in meaning of words or mathematics resulting from copy editor's rewriting.

- Missing mathematical symbols (e.g., $\eta \| Lx - b \|$ instead of $\eta = \| Lx - b \|$).

- Misplaced mathematical symbols or wrong kind: subscript should be superscript, $\mathbb{R}^{mxn}$ instead of $\mathbb{R}^{m \times n}$, A^t instead of A^T, $||A||$ instead of $\|A\|$.

- Errors in numbers in tables.

- Incorrect citation numbers. E.g., if a new reference [4] is added, every citation [n] must be renumbered to [n + 1] for $n \geq 4$, but this is easily overlooked.

Figure 8.2. Errors to check for when proofreading.

program listings include incorrect spacing and indentation and undesired line breaks. It is best to have a listing set from camera-ready copy, if possible.

Copy editors on mathematical journals often have mathematical qualifications, but they may still introduce errors in attempts to clarify, so read very carefully. For example, in one set of proofs that I received the phrase "stable in a sense weaker than" had been changed to "stable, and in a sense, weaker than", which altered the meaning. Since the original may have lacked clarity I changed the phrase to "stable in a weaker sense than". In every set of proofs I have received, I have found mistakes to correct. If you receive apparently perfect proofs, perhaps you are not proofreading carefully enough!

As an indication of how hard it can be to spot typographical errors I offer the following story. Abramowitz and Stegun's monumental *Handbook of Mathematical Functions* [2] was first published in 1964 and various corrections have been incorporated into later printings. As late as 1991, a new error was discovered: the right-hand side of equation (26.3.16) should read $Q(h) - P(k)$ instead of $P(h) - Q(k)$ [242]. This error was spotted when a Fortran subroutine from the NAG library that calculates probabilities behaved incorrectly; the error was traced to the incorrect formula in Abramowitz and Stegun's book.

I have spotted typographical errors in virtually every part of journal papers, including the title. The first sentence of one paper says that "One of the best unknown methods ... was developed by Jacobi." In one book preface the author thanks colleagues for helping him to produce a "final prodiuct" that is more accurate than he could have managed on his own. In the introduction of another, the author says that it would "be hard to underestimate" the importance of the subject of the book.

If you still need convincing of the importance of careful proofreading, consider the book on sky diving that was hurriedly recalled so that an erratum slip could be added. It read "On Page 8, line 7, 'State zip code' should read 'Pull rip cord'."

Corrections should be marked using the standard proofreading symbols; these will be listed on a sheet that accompanies the proofs. A list of proofreading symbols is given in Figure 8.3. In my experience it is possible to deviate slightly from the official conventions as long as the marking is clear. But make sure to mark corrections in the margin, which is the only place a printer looks when examining marked proofs; marks in the text should specify only the location of the correction. Take care to answer any queries raised by the copy editor (usually marked *Au.* in the margin). (The American Mathematical Society advises its editors to avoid asking "Is it this or that?" because many mathematicians are likely to answer "yes".) Some

Figure 8.3. Proofreading symbols.

journals require the printer's errors to be marked in one colour and the author's changes in another. Even if the copy editor doesn't ask, it is advisable to check the reference list to see if any unpublished or "to appear" references can be updated. Don't be afraid of writing notes to the copy editor (I often use yellow stick-on notes), particularly to praise their often under-appreciated work.

You should try to restrict changes you make on the proofs to corrections. If you make other changes you may have to pay for the extra typesetting costs. It is usually possible to add any vital, last-minute remarks (such as a mention of related work that has appeared since you submitted your manuscript) in a paragraph headed "Note Added in Proof" at the end of the paper.

At about the same time as you receive the proofs you will receive a reprint order form and, depending on the publisher, an invitation to pay page charges. Page charges cover the cost of typesetting the article and payment is usually optional for mathematics journals. Even if you request only the free reprints you still need to complete and return the reprint order form.

8.7. Author-Typeset TeX

I now focus on papers that are typeset in TeX by the author (TeX itself is discussed in Chapter 13). Many journals provide macros for use with TeX, LaTeX or AMS-LaTeX that produce output in the style of the journal. These are usually available from the journal Web page or by electronic mail from the editors or publishers.

In the same way that a computer programmer can write programs that are difficult to understand, an author can produce TeX source that is badly structured and contains esoteric macros, even though it is syntactically correct. Such TeX source is difficult to modify and this can lead to errors being introduced. If you intend to provide TeX source you should try to make it understandable. Watch out for precarious comments, such as those in the following example (the % symbol in TeX signifies that the rest of the line is a comment and should not be printed).

```
The widely used IEEE standard arithmetic
has $\beta=2$. % ANSI/IEEE Standard 754-1985
Double precision has $t=53$, $L=-1021$, $U=1024$,
and $u = 2^{-53} \approx 1.11 \times 10^{-16}$.
% IEEE arithmetic uses round to even.
```

This produces the output

> The widely used IEEE standard arithmetic has $\beta = 2$. Double precision has $t = 53$, $L = -1021$, $U = 1024$, and $u = 2^{-53} \approx 1.11 \times 10^{-16}$.

Suppose that, as these lines are edited on the computer, they are reformatted (either automatically, or upon the user giving a reformatting command) to give

```
The widely used IEEE standard arithmetic has
$\beta=2$. % ANSI/IEEE Standard 754-1985 Double
precision has $t=53$, $L=-1021$, $U=1024$, and
$u = 2^{-53} \approx 1.11 \times 10^{-16}$. % IEEE
arithmetic uses round to even.
```

As the comment symbols now act on different text, this results in the incorrect output

> The widely used IEEE standard arithmetic has $\beta = 2$. precision has $t = 53$, $L = -1021$, $U = 1024$, and $u = 2^{-53} \approx 1.11 \times 10^{-16}$. arithmetic uses round to even.

Because of this possibility I try to keep comments in a separate paragraph. So I would format the above example as

```
The widely used IEEE standard arithmetic has $\beta=2$.
Double precision has $t=53$, $L=-1021$, $U=1024$,
and $u = 2^{-53} \approx 1.11 \times 10^{-16}$.

% ANSI/IEEE Standard 754-1985
% IEEE arithmetic uses round to even.
```

Similar difficulties may arise if you edit with a line length of more than 80 characters (the standard screen width on many computers). A different text editor might wrap characters past the 80th position onto new lines, or, worse, truncate the lines; in either case, the meaning of the TeX source is changed.

Errors can be introduced in the transmission of TeX source by email. Characters such as ^, ~, {, } may be interchanged because of incompatibilities between the ASCII character set and other character sets used by certain computers. To warn of translation you could include test lines such as

```
%%   Exclamation    \!   Double quote  \"   Hash (number) \#
%%   Dollar          \$   Percent        \%   Ampersand      \&
%%   Acute accent    \'   Left paren    \(   Right paren    \)
```

at the top of your file.

Some mail systems object to lines longer than 72 characters. Some interpret a line beginning with the word *from* as being part of a mail message header and corrupt the line. Thus

```
from which it follows that
```

might be converted to

```
>from: which it follows that
```

which would be printed in TEX as

> ¿from: which it follows that

Another possible problem arises when ASCII files are transferred between Unix machines and DOS and Windows machines, since Unix terminates lines with a line-feed character, whereas DOS and Windows use a carriage-return line-feed pair. Public domain utilities, with names such as **dos2unix** and **unix2dos**, are available for converting between one format and the other.

The conclusions from this discussion are twofold.

(1) You should be aware of the potential problems and guard against them. Limit lines to 80 characters (or 72 characters if you will be mailing the file and are unsure which mail systems will be used), keep comment lines separate from the main text, and prepare source that is easy to read and understand.

(2) Read proofs of papers that are typeset from your source with just as much care as those that are re-typeset. Between submission and printing many errors can, potentially, be introduced.

If your paper is prepared in LATEX and you use nonstandard packages, make sure that you send the packages with the source. If you wish to avoid sending multiple files (which can be inconvenient by email), you can use the **filecontents** environment to put everything in one LATEX file. Suppose your paper uses the package **path**. Then you can insert

```
\begin{filecontents}{path.sty}
contents of path.sty
\end{filecontents}
```

before the \documentclass command. When the file is run through LATEX, the file **path.sty** is created if it does not exist; otherwise, a warning message is printed. Any number of **filecontents** environments can be included.

8.8. Copyright Issues

The responsibility for obtaining permission to reproduce copyrighted material rests with the author, rather than the publisher.

If you want to quote more than about 250 words from a copyrighted source, you need to obtain permission. Copyright law is vague about the exact number of words allowed; if in doubt, obtain permission. Permission is also required for previously published figures and tables if they are to be reproduced in their original form. For substantially altered figures or tables or paraphrased quoted material, permission is not required but a citation of the original source must be included.

All permissions must be obtained before a manuscript is submitted for publication. Write to the copyright owners (usually the publisher) with complete information about the material you would like to borrow and complete information about the book or paper you are writing. Specify whether you are applying for permission for print publication or electronic publication or both. It is usually helpful to send a duplicate letter since many copyright owners will grant permission directly on the letter and will retain the extra copy. In most cases permission is granted. Note that new permissions are usually required for new editions of a book.

To be safe, always cite the original source of any material reproduced from another publication, whether or not permission is necessary. It is also wise to check with your publisher (or check your contract) if you wish to reproduce material from your own previously published material.

When quoting text from another source, a convenient way to credit that source is to include publication information along with the quote, rather than in a footnote. When acknowledging permission to reproduce figures or tables, however, include the permission and original sources in the caption or a footnote. Some publishers may require exact credit lines; be sure to follow their instructions word for word.

8.9. A SIAM Journal Article—From Acceptance to Publication

What happens once your paper has been accepted for publication in one of SIAM's journals? The following description answers this question. You may be surprised at how many times your paper is checked for errors! This description also explains why you should not expect to be able to make late changes to the paper after you have dealt with the proofs. Note that this description may not be typical of the processes used by other publishers, especially those less involved with electronic publication.

TEX Papers

When the editor's acceptance letter arrives at the SIAM publications office, an editorial assistant processes the acceptance and sends acceptance correspondence to the author. Acceptance correspondence includes a request that the author send the TEX source file for the paper to SIAM immediately, or immediately after the SIAM style macros have been applied. Once received, the TEX file is "pre-TEXed" (macro verification and minor editorial changes). The electronic art is included, any non-electronic art is scanned, and the paper is printed out again. This new printout is then copy edited. After the paper is edited it goes to a TEX compositor (who may or may not be an in-house SIAM staffperson) for correction of the TEX file: the editing changes are made to the file, the file is re-run, and "first proofs" are printed. The paper may be proofread at this point. If many corrections are still necessary, a second round of "first pass" corrections may be done and new "first proofs" printed. Proofreading by the SIAM staff consists of checking the edited manuscript against the proofs to ensure that the requested edits were correctly incorporated into the file. The paper is not re-read. In effect, except for author corrections, the paper is completed at this point. First proofs are then mailed to the author, who is asked to return the marked proofs or a list of changes within 48 hours of receipt. Email or faxing of changes is encouraged.

After the author's changes are returned to SIAM, a SIAM staffperson incorporates the necessary corrections into the TEX file, the file is re-run, and "second proofs" are printed. These are proofed by a SIAM staffperson. As long as no typesetting or major editorial errors are found, the text of the paper is considered final at this point.

SIAM's on-line services manager assigns the paper to the issue of the journal that is currently being filled. The volume, issue, year, and page numbers are added to the TEX file. (The year is the year of electronic publication, which is the definitive publication date; the printed volume may appear in a later year.) Then the final PostScript file is generated, as well as `dvi` and PDF files. Finally, the files are published (posted on SIAM's Web server) as part of SIAM Journals Online (`http://epubs.siam.org`), in the appropriate issue. Issues are filled according to the print pagination budget and are (subsequently) printed and mailed according to the same budget.

Non-TEX Papers

When the editorial assistant sends acceptance correspondence to an author who has not already indicated that TEX source is available for the paper, the author is asked to confirm whether or not a TEX file is available. If there is no TEX file, the paper is copy edited and sent to a TEX compositor to be keyed into LATEX. The compositor prints the corrected paper ("first proofs"); the proofs are checked against the edited manuscript and sent to the author for proofreading. At this time the compositor sends the LATEX files to SIAM, where the remainder of the production process will be completed. The author is asked to return corrections to SIAM within 48 hours of receipt of the proofs. The rest of the process is the same as for TEX papers.

A Brief History of Scholarly Publishing (extract)

50,000 B.C. *Stone Age publisher demands that all manuscripts be double-spaced, and hacked on one side of stone only.*

1483 *Invention of* ibid.

1507 *First use of circumlocution.*

1859 *"Without whom" is used for the first time in list of acknowledgments.*

1916 *First successful divorce case based on failure of author to thank his wife, in the foreword of his book, for typing the manuscript.*

1928 *Early use of ambiguous rejection letter, beginning, "While we have many good things to say about your manuscript, we feel that we are not now in position . . . "*

1962 *Copy editors' anthem "Revise or Delete" is first sung at national convention. Quarrel over hyphenation in second stanza delays official acceptance.*

— DONALD D. JACKSON, in *Science with a Smile* (1992)

Publication Peculiarities

What is the record for the greatest number of authors of a refereed paper, where the authors are listed on the title page? My nomination is

> P. Aarnio et al. Study of hadronic decays of the Z^0 boson. *Physics Letters B*, 240(1,2):271–282, 1990. DELPHI Collaboration.

This paper has 547 authors from 29 institutions. The list of authors and their addresses occupies three journal pages. Papers with over 1,000 authors almost certainly exist, but the authors of these monster collaborations are usually listed in an appendix. For the shortest titles, I offer

> Charles A. McCarthy. c_p. *Israel J. Math.*, 5:249–271, 1967.
>
> Norman G. Meyers and James Serrin. $H = W$. *Proc. Natl. Acad. Sciences USA*, 51:1055–1056, 1964.

The latter title has the virtue of forming a complete sentence with subject, verb and object. The only title I have seen that contains the word OK is

> Thomas F. Fairgrieve and Allan D. Jepson. O.K. Floquet multipliers. *SIAM J. Numer. Anal.*, 28(5):1446–1462, 1991.

The next paper is famous in physics. Dyson explains that "Bethe had nothing to do with the writing of the paper but allowed his name to be put on it to fill the gap between Alpher and Gamow" [75].

> R. A. Alpher, H. Bethe, and G. Gamow. The origin of chemical elements. *Physical Review*, 73(7):803–804, 1948.

The only paper I know that has an animal as coauthor is

> J. H. Hetherington and F. D. C. Willard. Two-, three-, and four-atom exchange effects in bcc ^{3}He. *Physical Review Letters*, 35(21):1442–1443, 1975.

The story of how his cat (Felix domesticus) Chester, sired by Willard, came to be a coauthor is described by Hetherington in [291, pp. 110–111].

Chapter 9
Writing and Defending a Thesis

1987: Student writes Ph.D. thesis completely in verbatim *environment.*
— DAVID F. GRIFFITHS and DESMOND J. HIGHAM,
*Great Moments in L*A*T*E*X History*[15] (1997)

Calvin: I think we've got enough information now, don't you?
Hobbes: All we have is one "fact" you made up.
Calvin: That's plenty. By the time we add an introduction,
a few illustrations, and a conclusion,
it will look like a graduate thesis.
— CALVIN, *Calvin and Hobbes by Bill Watterson* (1991)

Remember to begin by writing the easiest parts first...
It is surprising how many people believe that a thesis...
should be written in the order
that it will be printed and subsequently read.
— ESTELLE M. PHILLIPS and D. S. PUGH, *How to Get a PhD* (1994)

---

[15]In [118].

Virtually all that has been said in Chapters 6 and 7 about writing and revising a paper applies to theses. (The term "dissertation" is synonymous with thesis, and is preferred by some.) In this chapter I give some specific advice that takes into account the special nature and purposes of a thesis written for an advanced degree. Much of the discussion applies to undergraduate level projects too.

9.1. The Purpose of a Thesis

The purpose of a thesis varies with the type of degree (Masters or Ph.D.) and the institution. The thesis might have to satisfy one or more of the following criteria:

- show that the student has read and understood a body of research literature,

- provide evidence that the student *is capable* of carrying out original research,

- show that the student has carried out original research,

- represent a significant contribution to the field.

It is worth checking what is expected by your institution.

9.2. Content

A thesis differs from a paper in several ways.

1. A thesis must be self-contained. Whereas a paper may direct the reader to another reference for details of a method, experimental results, or further analysis of a problem, a thesis must stand on its own as a complete account of the author's work on the subject of investigation.

2. A thesis is formatted like a book, broken into chapters rather than sections.

3. A thesis may include more than one topic, whereas a paper usually focuses on one.

4. A thesis is usually longer than an average paper, making good organization particularly important.

5. The primary readers of a thesis (possibly the only readers) are its judges, and they will read it with at least as much care as do the referees of a paper.

Since there is less pressure to save space than when writing a paper for a journal, you should generally include details in a thesis. It is important to demonstrate understanding of the subject, and phrases such as "it is easily shown that" and "we omit the proof" used in the presentation of original results may seem suspicious when you have no track record in the subject (the examiners may, of course, ask for such gaps to be filled during the oral examination). Trying to anticipate the examiners' questions should help you to decide what and how much to say on each topic.

The thesis should not be padded with unnecessary material (many theses are too long), but results that would not normally be published can be included, perhaps in an appendix, either because they might be of use to future workers or because you might want to refer to them in a paper based on the thesis. There is no "ideal" number of pages beyond which a thesis gains respectability, and indeed there is great variation in the length of theses among different subjects and even within a subject. The supervisor (UK) or thesis advisor (US) can offer advice about the suitable length.

A thesis has a fairly rigid structure. In the first one or two chapters the problem being addressed must be clearly described and put into context. You are expected to demonstrate a sound knowledge and understanding of the existing work on the topic by providing a critical survey of the relevant literature. If there is more than one possible approach to the problem, the choice of method must be justified. For a computational project the method developed or investigated in the thesis would normally be compared experimentally with the major alternatives.

At the end of the thesis, conclusions must be carefully drawn and the overall contribution of the thesis assessed. It is a good idea to identify open problems and future directions for research, since being able to do so is one of the attributes required of a researcher. Note that a thesis does not necessarily have to present major new or improved results; in many cases the key requirement is the development and communication of original ideas using sound techniques.

When you write a thesis you are usually relatively inexperienced at technical writing, so it is important to avoid inadvertently committing plagiarism (see §6.16). If you copy text word for word from another source you must put it in quotation marks and cite the source. If you find yourself copying, or paraphrasing, someone else's proof of a theorem, ask yourself if you need to give the proof—if it is not your own work, will it add anything to the thesis? Examiners will be particularly alert to the possibility of plagiarism, so be careful to avoid committing this sin.

9.3. Presentation

Each institution has rules about the presentation of a thesis. Page, font and margin sizes, line spacing (often required to be double or one and a half times the standard spacing) and the form of binding may all be tightly regulated, and non-conforming theses may be rejected on submission. The opening pages will be required to follow a standard format, typically comprising the following items (some of which will be optional).

1. A title page, listing the author, title, department, type of degree, and year (and possibly month) of submission.

2. A declaration that the work has not been used in another degree submission.

3. A statement on copyright and the ownership of intellectual property rights.

4. A list of notation.

5. A brief statement of the author's research career.

6. Acknowledgements and dedications.

7. Table of contents.

8. List of figures.

9. List of tables.

10. Abstract. The abstract may need to be repeated on a separate application form. Once the degree has been obtained, the abstract is likely to be entered into a database such as *Dissertation Abstracts International* or *Index to Theses* (for theses from universities in Great Britain and Ireland, and available on the Web for registered users at `http://www.theses.com`).

The opening pages are also the place to indicate which parts of the thesis (if any) have already been published, and which parts are joint work.

Your library will contain previous successful theses, which you can inspect to check the required format—but bear in mind that rules of presentation can and do change. It is likely that a LaTeX package will be available at your institution for typesetting theses in the official style.

I recommend producing an index for the thesis, although this practice is not common. A well-prepared index (see §13.4) can be a significant aid to examiners and readers.

When should you start to write the thesis? My advice is to start sooner rather than later. In the early months of study in which you become familiar with the problem and the literature you can begin to draft the first few chapters. You should also start immediately to collect references for the bibliography—it is difficult in the later stages to hunt for half-remembered references. One reason for making an early start on writing the background and survey material is that at this stage you will be enthusiastic—later, you may know this material so well that it seems dull and boring.

I encourage my students to write up their work in LaTeX as they progress through the period of study, so that when the time comes to produce the thesis much of the writing has been done. Since most students now typeset their own theses, this approach allows them to learn the typesetting system when they are least stressed, rather than in the last hectic months. If the thesis work has progressed rapidly, they may be in the pleasant position of having one or more papers already written, upon which they can base the thesis.

Unlike a published paper, a thesis will not be read by a copy editor or proofreader. It is therefore particularly important that you thoroughly read and check the thesis before it is submitted. Your thesis advisor should read and comment on the thesis, and it is worthwhile recruiting fellow students as readers, too; even if they are not specialists in the area they should be able to offer useful suggestions for improvement.

9.4. The Thesis Defence

The oral defence of a thesis takes different forms in different countries. For example, in the UK the candidate answers questions posed by the examiners, but does not usually give a formal presentation, whereas in other European countries it is more common for the candidate to give a presentation followed by questions from the jury. The number of examiners also varies greatly between countries.

Perhaps the most important piece of advice applicable to all forms of defence is to read the thesis beforehand. The defence may take place weeks or months after you submitted the thesis, and in the meantime you may forget exactly what material you included in the thesis and where it is located. To be properly prepared you need to know the thesis inside out.

One of the purposes of the defence is for the examiners to satisfy themselves that you (not someone else) did the work you claim to have done and that you understand it. As well as asking straightforward questions about the thesis, they may therefore ask you why you took the approaches you did and to justify assumptions and amplify arguments. You can also expect questions that explore your knowledge of the literature outside the immediate area of the thesis, as the examiners gauge your general familiarity with the research area.

It is important that you listen to questions carefully and answer the question that is asked, not some other question. When you are under pressure it is easy to misunderstand what the examiners ask you. If you do not understand a question, say so, and the question will be repeated or rephrased.

If you give a formal presentation (typically 40–50 minutes long) you should aim to give an overview of the research area and the work you have done and not to go too deeply into the details. The examiners will want to see that you appreciate the context and significance of your work and that you are aware of problems remaining for future research. Consult Chapters 10 and 11 for practical advice on writing and giving the talk.

Finally, note that an examiner who has carefully read your thesis and attended the defence should know enough about you to write a reference for your job applications. The examiners may even be able to offer advice on where to seek employment.

Oral Examination Procedure (extract)

1. Before beginning the examination, make it clear to the examinee that his whole professional career may turn on his performance. Stress the importance and formality of the occasion. Put him in his proper place at the outset.

2. Throw out your hardest question first. (This is very important. If your first question is sufficiently difficult or involved, he will be too rattled to answer subsequent questions, no matter how simple they may be.) . . .

9. Every few minutes, ask him if he is nervous . . .

11. Wear dark glasses. Inscrutability is unnerving.

12 Terminate the examination by telling the examinee, "Don't call us; we will call you."

— S. D. MASON, in *A Random Walk in Science* (1973)

9.5. Further Reading

Books are available that give advice on the whole process of doing research for a Ph.D., including writing the thesis. For a UK perspective, Phillips and Pugh [226] can be recommended, while Sternberg [258] and Rudestam and Newton [241] offer a US viewpoint.

Chapter 10
Writing a Talk

My recommendations amount to this ...
Make your lecture simple (special and concrete);
be sure to prove something and ask something;
prepare, in detail;
organize the content and adjust to the level of the audience;
keep it short, and, to be sure of doing so,
prepare it so as to make it flexible.

— PAUL R. HALMOS, *How to Talk Mathematics* (1974)

I always find myself obliged,
if my argument is of the least importance,
to draw up a plan of it on paper and
fill in the parts by recalling them to mind,
either by association or otherwise.

— MICHAEL FARADAY[16]

An awful slide is one which contains approximately a million numbers
(and we've left our opera glasses behind).
An awful lecture slide is one which shows a complete set of
engineering drawings and specifications for a super-tanker.

— KODAK LIMITED, *Let's Stamp Out Awful Lecture Slides* (1979)

---

[16]Quoted in [271, p. 98].

10.1. What Is a Talk?

In this chapter I discuss how to write a mathematical talk. By talk I mean
a formal presentation that is prepared in advance, such as a departmental
seminar or a conference talk, but not one of a series of lectures to students.
In most talks the speaker writes on a blackboard or displays pre-written
transparencies on an overhead projector. (From here on I will refer to
transparencies as slides, since this term is frequently used and is easier to
write and say.) I will restrict my attention to slides, which are the medium
of choice for most speakers at conferences, but much of what follows is
applicable to the blackboard. I will assume that the speaker uses the slides
as a guide and speaks freely. Reading a talk word for word from the slides
should be avoided; one of the few situations where it may be necessary is if
you have to give a talk in a foreign language with which you are unfamiliar
(Kenny [149] offers some advice on how to do this).

A talk has several advantages over a written paper [50].

1. Understanding can be conveyed in ways that would be considered too
   simplified or lacking in rigour for a journal paper.

2. Unfinished work, or negative results that might never be published,
   can be described.

3. Views based on personal experience are particularly effective in a talk.

4. Ideas, predictions and conjectures that you would hesitate to commit
   to paper can be explained and useful feedback obtained from the
   audience.

5. A talk is unique to you—no one else could give it in exactly the same
   way. A talk carries your personal stamp more strongly than a paper.

Given these advantages, and the way in which written information is
communicated in a talk, it is not surprising that writing slides differs from
writing a paper in several respects.

1. Usually, less material can be covered in a talk than in a corresponding
   paper, and fewer details need to be given.

2. Particular care must be taken to explain and reinforce meaning, notation and direction, for a listener is unable to pause, review what
   has gone before, or scan ahead to see what is coming.

3. Some of the usual rules of writing can be ignored in the interest of
   rapid comprehension. For example, you can write non-sentences and
   use abbreviations and contractions freely.

4. Within reason, what you write can be imprecise and incomplete—
and even incorrect. These tactics are used to simplify the content of
a slide, and to avoid excessive detail. Of course, to make sure that no
confusion arises you must elaborate and explain the hidden or falsified
features as you talk through the slide.

10.2. Designing the Talk

The first step in writing a talk is to analyse the audience. Decide what
background material you can assume the listeners already know and what
material you will have to review. If you misjudge the listeners' knowledge,
they could find your talk incomprehensible at one extreme, or slow and
boring at the other. If you are unsure of the audience, prepare extra slides
that can be included or omitted depending on your impression as you go
through the talk and on any questions received.

The title of your talk should not necessarily be the same as the one
you would use for a paper, because your potential audience may be very
different from that for a paper. To encourage non-specialists to attend the
talk keep technical terms and jargon to a minimum. I once gave a talk
titled "Exploiting Fast Matrix Multiplication within the Level 3 BLAS" in
a context where non-experts in my area were among the potential audience.
I later found out that several people did not attend because they had not
heard of BLAS and thought they would not gain anything from the talk,
whereas the talk was designed to be understandable to them. A better title
would have been the more general "Exploiting Fast Matrix Multiplication
in Matrix Computations".

A controversial title that you would be reluctant to use for a paper may
be acceptable for a talk. It will help to attract an audience and you can
qualify your bold claims in the lecture. Make sure, though, that the content
lives up to the title.

It is advisable to begin with a slide containing your name and affiliation
and the title of your talk. This information may not be clearly or correctly enunciated when you are introduced, and it does no harm to show
it again. The title slide is an appropriate place to acknowledge co-authors
and financial support.

Because of the fixed path that a listener takes through a talk, the structure of a talk is more rigid than that of a paper. Most successful talks
follow the time-honoured format "Tell them what you are going to say, say
it, then tell them what you said." Therefore, at the start of the talk it is
usual to outline what you are going to say: summarize your objectives and
methods, and (perhaps) state your conclusions. This is often done with the
aid of an overview slide but it can also be done by speaking over the title

```
  • Introduction and Motivation
  ▷ Deriving Partitioned Algorithms
  • Block LU Factorization and Matrix Inversion
  • Exploiting Fast Matrix Multiplication
```

Figure 10.1. Contents slide: the triangle points to the next topic.

slide. The aim is to give the listeners a mental road-map of the talk. You also need to sprinkle signposts through the talk, so that the listeners know what is coming next and how far there is to go. This can be done orally (example: "Now, before presenting some numerical examples and my overall conclusions, I'll indicate how the result can be generalized to a wider class of problems"), or you can break the talk into sections, each with its own title slide. Another useful technique is to intersperse the talk with contents slides that are identical apart from a mark that highlights the topic to be discussed next; see Figure 10.1.[17]

Garver [102] recommends lightening a talk by building in multiple entry points, at any of which the listener can pick up the talk again after getting lost. An entry point might be a new topic, problem or method, or an application of an earlier result that does not require an understanding of the result's proof. Multiple exit points are also worth preparing if you are unsure about how the audience will react. They give you the option of omitting chunks of the talk without loss of continuity. A sure sign that you should exercise this option is if you see members of the audience looking at their watches, or, worse, tapping them to see if they have stopped!

An unusual practice worth considering is to give a printed handout to the audience. This might help the listeners to keep track of complicated definitions and results and save them taking notes, or it might give a list of references mentioned in the talk. A danger of this approach is that it may be seen as presuming the audience cannot take notes themselves and are interested enough in the work to want to take away a permanent record. Handouts can, alternatively, be made available for interested persons to pick up after the talk.

---

[17]To save space, all the boxes that surround the example slides in this chapter are made just tall enough to hold the slide's content.

10.3. Writing the Slides

Begin the talk by stating the problem, putting it into context, and motivating it. This initial scene-setting is particularly important since the audience may well contain people who are not experts in your area, or who are just beginning their research careers.

The most common mistake in writing a talk is to put too much on the individual slides. The maxim "less is more" is appropriate, because a busy, cluttered slide is hard for the audience to assimilate and may divert their attention from what you are saying. Since a slide is a visual aid, it should contain the core of what you want to say, but you can fill in the details and explanations as you talk through the slide. (If you merely read the slide, it could be argued that you might as well not be there!) There are various recommendations about how many lines of type a slide should contain: a maximum of 7–8 lines is recommended by Kenny [149] and a more liberal 8–10 lines by Freeman et al. [86]. These are laudable aims, but in mathematical talks speakers often use 20 or more lines, though not always to good effect.

A slide may be too long for two reasons: the content is too expansive and needs editing, or too many ideas are expressed. Try to limit each slide to one main idea or result. More than one may confuse the audience and weaken the impact of the points you try to make. A good habit is to put a title line at the top of each slide; if you find it hard to think of an appropriate title, the slide can probably be improved, perhaps by splitting it into two.

Don't present a detailed proof of a theorem, unless it is very short. It is far better to describe the ideas behind the proof and give just an outline. Most people go to a talk hoping to learn new ideas, and will read the paper if they want to see the details.

When a stream of development stretches over several slides, the audience might wish to refer back to an earlier slide from a later one. To prevent this you can replicate information (an important definition or lemma, say) from one slide to another. A related technique is to build up a slide gradually, by using overlays, or simply by making each slide in a series a superset of the previous one. (The latter effect can be achieved by covering the complete slide with a sheet of paper, and gradually revealing the contents; but be warned that many people find this peek-a-boo style irritating. I do not recommend this approach, but, if you must use it, cover the slide *before* it goes on the projector, not after.) Overlays are best handled by taping them together along one side, and flipping each one over in turn, since otherwise precise alignment is difficult.

If you think you will need to refer back to an earlier slide at some particular point, insert a duplicate slide. This avoids the need to search through the pile of used slides. It is worth finding out in advance whether two projectors will be available. If so, you will have less need to replicate material because you can display two slides at a time.

It is imperative to number your slides, so that you can keep them in order at all times. At the end of the talk the slides will inevitably be jumbled and numbers help you to find a particular slide for redisplaying in answer to questions. I put the number, the date of preparation, and a shortened form of the title of the talk, on the header line of each slide.

When you write a slide, aim for economy of words. Chop sentences mercilessly to leave the bare minimum that is readily comprehensible. Here are some illustrative examples.

Original: It can be shown that $\partial \|Ax\| = \{A^T \mathrm{dual}(Ax)\}$.

Shorter: Can show $\partial \|Ax\| = \{A^T \mathrm{dual}(Ax)\}$.

Even shorter: $\partial \|Ax\| = \{A^T \mathrm{dual}(Ax)\}$.

Original: The stepsize is stable if $p(z)$ is a Schur polynomial, that is, if all its roots are less than one in modulus.

Shorter: Stepsize is stable if $p(z)$ is a Schur polynomial (all roots less than one in modulus).

Even shorter: Stepsize stable if $p(z)$ is Schur ($|\mathrm{roots}| < 1$).

Original: We can perform a similar analysis for the Lagrange interpolant.

Shorter: Similar analysis works for the Lagrange interpolant.

Even shorter: OK for Lagrange.

Lists are an effective way of presenting information concisely. Audiences like them, because there is something psychologically appealing about being given a list of things to note or remember. Consider using bulleted or numbered lists of five or fewer items, expressed with parallel constructions where appropriate (see §4.16).

If you cite your own work you can reduce your name to one letter, but the names of others should not be abbreviated. Thus, from one of my slides:

Original: This question answered in nonsingular case by Higham and Knight (1991).

Shorter: Answered for nonsingular A by H & Knight (1991).

Be sure to avoid spelling mistakes on the slides. They are much more noticeable and embarrassing in slides than in a paper.

Graphs and pictures help to break up the monotony of wordy slides. But beware of "slide shuffling", a common mistake where a long sequence of similar slides is displayed in rapid succession ("that was the plot for $\rho = 0.5$, now here is $\rho = 0.6$, ..., and here is $\rho = 0.7$,..."). An audience needs time to absorb each slide and it is difficult for anyone to remember more than the last two slides. Anything that might irritate or bore your audience is to be avoided.

At the end of the talk, conclusions and a summary are usually given. This is your last chance to impress your ideas upon the audience. If possible, try to summarize what you have said in a way that does not presuppose understanding of the talk. Even if your conclusions are largely negative, try to finish the talk on a positive note! The last slide is a good place to give your email address and a Web address where your relevant papers can be found.

Legibility of the Slides

It is of paramount importance that slides be legible when they are projected. Unfortunately, slides are frequently difficult to read for people sitting towards the rear of the room.

An important point to appreciate is that many overhead projector installations exhibit keystoning, whereby the image is wider at the top than at the bottom; this is avoided if the screen is tilted forward. Keystoning causes characters at the bottom of the screen to look much smaller than those at the top.

Take care not to write too close to the edges of a slide, as the areas near the edges may not be clearly visible when the slide is projected. And try to leave a sizeable gap at the bottom of the slide, as the bottom of the screen may be obstructed from the view of people at the back of the audience. I recommend experimenting with a projector to determine acceptable limits.

Various recommendations have been given in the literature on the minimum height of characters, but they necessarily involve assumptions about the magnification of the original image. My advice is to experiment with a projector in an empty room of a similar size to the one that will be used for the talk—you can quickly determine how large the characters need to be. I find that

characters this large (\LARGE in LaTeX at 10pt)

produce a slide readable from the back of most lecture rooms. For the comfort of the audience I prefer to use

characters this large (\huge in LaTeX at 10pt).

For a given character size the font also affects readability [240, Chap. 2]. Some people prefer to use sans serif fonts for slides (fonts without small extensions at the ends of strokes):

sans serif font

It is not uncommon to see slides that have been photocopied from a paper without magnification. Invariably the slides are difficult or impossible to read. Avoid this sin; prepare concise, legible slides specially for the talk. The practice of photocopying from a paper is most common with tables. Better even than magnifying the original table is to rewrite it, pruning it to the information essential for that slide. For example, some entries in a table of numbers may be dispensable, and it may be possible to reduce the number of significant digits. More so than in a paper, a table of numerical results may be better displayed as a graph, which is easier to absorb as long as the labels are readable.

How Many Slides?

The number of slides needed depends on several factors, such as the length of the talk, the amount of mathematical detail to be presented, and the number of graphs and pictures. It also depends on the mode of presentation: a speaker who uses an overhead projector to display outline notes and who spends considerable time filling in the details on the blackboard will need far fewer slides than one who uses only the projector. In my experience, it is best to allow about two minutes per slide for "slide only" talks (the same recommendation is made in [86] and [102]). Allowing for variation in slide length, for a twenty minute talk you should aim for twelve slides or fewer, while for a fifty minute talk thirty slides is about the upper limit.

Handwritten or Typeset?

Which are better: handwritten or typeset slides? It is a matter of personal preference. The best handwritten slides are just as good as the best typeset ones, and the worst of both kinds are very poor. I produce all my slides in LaTeX, for several reasons. I can work from the LaTeX source for the corresponding paper, which saves a lot of typing; typeset slides can easily be adapted and reprinted for use in other talks; I find it easier to revise and correct typeset slides than handwritten ones, and this encourages me to produce better slides; and I have no worries about spilling liquid over my slides or losing them (they are never further away than an ftp call to my workstation). Advantages of handwritten slides are that they can be modified right up until the start of the talk without the need for access to a computer or printer, it is easy and inexpensive to use colours, and handwritten slides are more individual.

LaTeX 2_ε has a document class `slides` [172, §5.2] designed for making slides (it supersedes SLiTeX, a version of LaTeX 2.09). It automatically produces large characters in a specially designed sans serif font. It supports the production of coloured slides, via the `color` package, and overlays (so that one slide can be superimposed on another). Various other macro packages are available for use with LaTeX, including the `seminar` package by Timothy Van Zandt (available from CTAN—see §13.5). Note that typeset slides are easier to read if they are not right-justified and if hyphenation is turned off.

To break the monotony of black and white typeset slides you can underline headings and key phrases with a coloured pen. You can also leave gaps to be filled during the talk, later wiping the slides clean, ready for the next time you give the talk.

If you use typeset slides you may have the option of projecting directly from a computer through a projection device, as opposed to using transparencies on an overhead projector. One advantage of working from the computer is that you can change the slides right until the last minute. Another is that programs such as Microsoft PowerPoint permit sophisticated techniques to be used, such as animation and building up a slide layer by layer. There are several disadvantages:

- The quality of computer projection systems is variable. Unless a sharp, bright image is produced, it will be better to use transparencies.

- There is a tendency for authors to emphasize presentation at the expense of content when using the fancier presentation programs. One seminar organizer I know has banned the use of PowerPoint presentations.

- More things can (and do) go wrong so, as a golden rule, have printed slides ready just in case!

10.4. Example Slides

In this section I give some examples of how to improve slides. In each case two slides are shown: an original draft and a revised version. Keep in mind that the slides would be printed in much larger type, and magnified further by the projector. Therefore, what appear here to be only minor flaws will be much more noticeable in practice.

The slide in Figure 10.2 gives some numerical results. The second version improves on the first in several ways.

- The definition of the matrix has been changed from a precise but cryptic form to a descriptive one.

- The quantities ρ^N, ρ^R, and γ_2 would have been defined on an earlier slide, but the audience may need to be reminded of what they represent. Therefore, a description in words has been added.

- Since only the order of magnitude of the error is relevant, the numbers are quoted to just one significant figure. This reduces the visual complexity of the table. To reduce the complexity further, the column headed $\rho^C(\widehat{y})$ has been removed, and the minimum number of rules has been used in the table.

In Figure 10.3, the original slide contains a theorem as it would be stated in a paper. The revised slide omits some of the conditions, which can be mentioned briefly as the slide is explained, and reminds the viewer of the definition of $\kappa_2(A)$. The simplified slide is much easier to assimilate.

The revised slide in Figure 10.5 omits some unnecessary words and symbols from the slide in Figure 10.4. The slide in Figure 10.6 has far too many words. Ruthless pruning produces the much more acceptable version in Figure 10.7; as usual, the information removed from the slide would be conveyed when giving the talk. Note that the parentheses in the expressions for β_{n+1} and π_{n+1} were unnecessary and so have been omitted, and the final error bound is written in a more readable way. Another possible improvement is to remove the definitions of α_0, β_0 and π_0, which may be irrelevant for this slide.

10.5. Further Reading

See §11.3 in the next chapter.

Example

The unit roundoff is $u \approx 1.1921 \times 10^{-7}$.

$$A = \mathrm{randsvd}([10, 16], 1\mathrm{e}2)$$

$$\kappa_2(A) = 1.00\mathrm{e}2, \ \mathrm{cond}_2(A) = 8.63\mathrm{e}1$$
$$\mathrm{cond}_2(A, x) = 1.57\mathrm{e}2$$

	$\rho^N(\widehat{y})$	$\rho^R(\widehat{y})$	$\rho^C(\widehat{y})$	$\gamma_2(\widehat{y})$
Q method	1.83e−8	9.88e−9	1.42e−7	2.01e−6
SNE	5.11e−7	2.79e−7	4.40e−6	4.97e−6
	1.52e−8	6.45e−9	9.64e−8	1.99e−6

Example

Unit roundoff $u \approx 1.2 \times 10^{-7}$.

$$A \quad 10 \times 16, \quad \mathrm{random}$$

$$\kappa_2(A) = 1\mathrm{e}2, \quad \mathrm{cond}_2(A) = 9\mathrm{e}1$$

	ρ^N	ρ^R	γ_2
Q method	2e−8	1e−8	2e−6
SNE	5e−7	3e−7	5e−6
	2e−8	6e−9	2e−6

ρ^N = normwise backward error

ρ^R = row-wise backward error

γ_2 = forward error

Figure 10.2.

PERTURBATION BOUNDS

Traditional Result

Theorem Let $A \in \mathbb{R}^{m \times n}$ and $0 \neq b \in \mathbb{R}^m$. Suppose
$\text{rank}(A) = m \leq n$ and that $\Delta A \in \mathbb{R}^{m \times n}$ and $\Delta b \in \mathbb{R}^m$ satisfy

$$\epsilon = \max\{\|\Delta A\|_2/\|A\|_2, \|\Delta b\|_2/\|b\|_2\} < \sigma_m(A).$$

If x and $\widehat{x}$ are the minimum norm solutions to $Ax = b$ and
$(A + \Delta A)\widehat{x} = b + \Delta b$ respectively, then

$$\frac{\|\widehat{x} - x\|_2}{\|x\|_2} \leq \min\{3, n - m + 2\}\kappa_2(A)\epsilon + O(\epsilon^2).$$

PERTURBATION BOUNDS

$A \in \mathbb{R}^{m \times n}$, $m \leq n$.

$$x = \text{ min norm sol'n to } Ax = b,$$
$$\widehat{x} = \text{ min norm sol'n to } (A + \Delta A)\widehat{x} = (b + \Delta b).$$

Traditional Bound

$$\frac{\|\widehat{x} - x\|_2}{\|x\|_2} \leq c_{m,n}\, \kappa_2(A)\epsilon + O(\epsilon^2),$$

where

$$\epsilon = \max\left\{\frac{\|\Delta A\|_2}{\|A\|_2}, \frac{\|\Delta b\|_2}{\|b\|_2}\right\},$$

$$\kappa_2(A) = \|A^+\|_2\|A\|_2.$$

Figure 10.3.

MACHIN'S ARCTAN FORMULA

John Machin (1680–1752) used Gregory's arctan series to obtain a more rapidly convergent series for computing π. He argued as follows.

If $\tan\theta = 1/5$, then

$$\tan 2\theta = \frac{2\tan\theta}{1 - \tan^2\theta} = \frac{5}{12}$$

and

$$\tan 4\theta = \frac{2\tan 2\theta}{1 - \tan^2 2\theta} = \frac{120}{119}.$$

Since $\tan 4\theta = 120/119$ and $\tan(\pi/4) = 1$, this suggests we evaluate

$$\tan(4\theta - \pi/4) = \frac{\tan 4\theta - 1}{1 + \tan 4\theta} = \frac{1}{239}.$$

This gives

$$\arctan\frac{1}{239} = 4\theta - \frac{\pi}{4}$$

$$= 4\arctan\frac{1}{5} - \frac{\pi}{4},$$

or

$$\pi = 16\arctan\frac{1}{5} - 4\arctan\frac{1}{239}.$$

In 1706 Machin used this formula to compute 100 digits of π.

Figure 10.4.

MACHIN'S ARCTAN FORMULA

John Machin (1680–1752): if $\tan \theta = 1/5$, then

$$\tan 2\theta = \frac{2 \tan \theta}{1 - \tan^2 \theta} = \frac{5}{12}.$$

Similarly, $\tan 4\theta = 120/119$. Then

$$\tan(4\theta - \pi/4) = \frac{\tan 4\theta - 1}{1 + \tan 4\theta} = \frac{1}{239}$$

$$\Rightarrow \quad \arctan \frac{1}{239} = 4\theta - \frac{\pi}{4}$$

$$= 4 \arctan \frac{1}{5} - \frac{\pi}{4}.$$

In 1706 Machin used Gregory's arctan series to compute 100 digits of π.

Figure 10.5.

Fast Methods for Computing π

Up to the 1970s all calculations of π were done using linearly convergent methods based on series, products, or continued fractions. For these, doubling the number of digits in the result roughly doubles the work.

Brent and Salamin independently discovered a quadratically convergent iteration for π in 1976. For such an iteration, the errors satisfy $e_{k+1} = O(e_k^2)$, so each iteration step approximately doubles the number of correct digits.

Borwein and Borwein (1984) discovered another quadratically convergent iteration, and later found even higher order iterations. Their 1984 iteration is

$$\alpha_0 = \sqrt{2}, \quad \beta_0 = 0, \quad \pi_0 = 2 + \sqrt{2},$$

$$\alpha_{n+1} = \frac{1}{2}(\alpha_n^{1/2} + \alpha_n^{-1/2}),$$

$$\beta_{n+1} = \alpha_n^{1/2} \left(\frac{\beta_n + 1}{\beta_n + \alpha_n} \right),$$

$$\pi_{n+1} = \pi_n \beta_{n+1} \left(\frac{1 + \alpha_{n+1}}{1 + \beta_{n+1}} \right),$$

for which $|\pi_n - \pi| \leq \frac{1}{10^{2^n}}$.

Figure 10.6.

Fast Methods for π

Up to 1970s: linearly convergent series, products, or continued fractions.

Brent and Salamin (independently) 1976: quadratically convergent iteration for π. Errors satisfy $e_{k+1} = O(e_k^2)$.

Borwein and Borwein (1984, 1987): another quadratically convergent iteration and higher order ones. Quadratic iteration: $\alpha_0 = \sqrt{2}$, $\beta_0 = 0$, $\pi_0 = 2 + \sqrt{2}$,

$$\alpha_{n+1} = \tfrac{1}{2}(\alpha_n^{1/2} + \alpha_n^{-1/2}),$$

$$\beta_{n+1} = \alpha_n^{1/2} \frac{\beta_n + 1}{\beta_n + \alpha_n}, \quad \pi_{n+1} = \pi_n \beta_{n+1} \frac{1 + \alpha_{n+1}}{1 + \beta_{n+1}},$$

for which $|\pi_n - \pi| \leq 10^{-2^n}$.

Figure 10.7.

Chapter 11
Giving a Talk

You are now ready to start your speech...
Do not thank the Chairman or thank the audience.
Do not attempt to tell them that you are pleased to be there,
and above all do not elaborate on your lack of skill as a public speaker.
Audiences do not like speakers to apologise.
They find it embarrassing.

— PETER KENNY, *A Handbook of Public Speaking*
for Scientists and Engineers (1982)

The utterance should not be rapid and hurried and consequently unintelligible
but slow and deliberate
conveying ideas with ease from the Lecturer and
infusing them with clearness and readiness
into the minds of the audience.

— MICHAEL FARADAY[18], *Letter to B. Abbott* (11 June 1813)

A Lecturer falls deeply beneath the dignity of his character
when he descends so low as to angle for claps
and asks for commendation
yet have I seen a lecturer even at this point.

— MICHAEL FARADAY[18], *Letter to B. Abbott* (18 June 1813)

The Three Rules of Public Speaking:
Be forthright.
Be brief.
Be seated.

— SUSAN DRESSEL AND JOE CHEW, *Authenticity Beats Eloquence* (1987)

---

[18]In [297].

The previous chapter described how to design a talk and prepare the slides. Now I turn to the process of giving the talk.

11.1. Preparation

Having written the slides, you need to decide what you want to say while each slide is displayed. You should explain and amplify what is on the slides rather than simply read them word for word, and add anecdotes, stories and humour as you feel appropriate. If you are an inexperienced speaker you should find it helpful to begin by writing the talk out in full sentences, taking care to include everything that you want to say. This text should not be used for the actual presentation, since reading from a prepared text always sounds unnatural and stilted, but should be the basis for preparing notes from which you work. I like to write on the backing paper for each slide the key words and phrases that I will use to introduce and explain the slide. I use these notes when practising, but I find that I usually know them by heart by the time of the talk. If I give the talk again a few weeks or months later, the notes save me a lot of preparation time. The notes can alternatively be written onto prompt slides inserted between the real slides and numbered specially (1a, 2a, etc.). A few of the prompt slides can even be displayed on the projector if the occasion demands it. Whatever approach you use, try to use the initial period of planning to make your language more vivid by finding synonyms for frequently occurring words and developing colourful imagery.

Most experienced speakers are happy if the audience go away having understood and remembered one point or idea from the talk. You should therefore try to emphasize just one or two key points, repeating them in different words and at different points in the talk, to ensure that they are taken in.

Make sure you know how to pronounce all the technical terms and names that you will use in the talk, since mispronunciation can be embarrassing for both you and the audience.

Obviously you should practise the talk, but how much practice is needed varies from person to person and also with experience. I need to practise my talk, speaking aloud, two or three times before I am happy with it, but I find that too much practice can make the talk too slick and lacking in spontaneity. It is important to time each practice talk, so that you can adjust the length to avoid overrunning or finishing very early. For an inexperienced speaker, it is very helpful to give a practice talk in front of friends and colleagues, especially if they can be persuaded to offer constructive criticism.

Needless to say, it is vital to ensure that your slides have been correctly prepared and that you do not lose them on the way to the venue. This way, you should avoid the embarrassment of the professor who began his talk at a major conference only to realise that his secretary had photocopied onto the paper backing sheets and not the slides themselves!

On the day of the talk or earlier, I strongly recommend going in advance to the room where the talk will be given and investigating the following questions.

- How big is the room? How loud will you have to speak to be heard? (Remember that a voice sounds louder in an empty room, because bodies absorb sound.) Will a microphone be used? If so, it will almost certainly be a clip-on type, possibly with a cigarette packet-sized unit that you must put in a pocket or clip onto a belt; make sure you are suitably attired. A hand-held microphone, occasionally encountered, is difficult to hold at a constant distance from the mouth while you move around, and makes it difficult to talk while you change slides.

- What pointing strategy will you use: to point at the screen or at the slide on the projector? The screen may be too big or far away to point at with a traditional metal or wooden pointer, though a laser pointer may still be usable. Laser pointers are difficult to use effectively: they need to be held very steady to avoid shake and the unpredictable, short-lived appearances of the (usually red) dot at different points on the slide can irritate the audience. If you do point at the screen, take care not to talk into the screen, or your voice will not project into the audience. If you point at the slide on the projector use a pen or pointer rather than a finger, and lay it on the projector if possible to avoid shake.

- Where will you put your pile of slides? Usually, there is a suitable table close to the projector. The location of the table and the projector will determine where you stand as you give the talk. Consider whether it is possible to move them to a better position. Try not to walk across the projector's image, though this is often unavoidable if you are using two projectors.

- Is the projector working and do you know how to use it? Find the on/off switch for the projector (sometimes a nontrivial task!). Put a slide on the projector and check that the projector is correctly aligned and sharply focused. Check if there is a spare bulb and how it is accessed. Find a place to stand where you will not block the audience's view of the screen. If there are two projectors, compare the clarity and size of their images, and if one is clearly superior consider using it as the master projector rather than alternating between the two. Put a transparency on the projector and see if it slips; if so, it will need holding in place with a coin or eraser (or the projector can be propped up with a folded piece of paper under its legs).

- Where are the light switches? The chairperson should take care of extinguishing the lights near the projector screen, but you might have to step in if he or she is unprepared.

- Where is the chalk (if there is a blackboard), or where are the pens (if there is a whiteboard)? You may want to write on the board to explain something more fully, or in answer to a question. You may need to correct or add to a slide, so have your pens handy.

This reconnaissance not only prepares you for most eventualities, but makes you feel more confident in the run-up to the talk because you *know* that you are prepared.

On the day of the talk, if you are not the first speaker you have an advantage in that you can observe the other speakers and note and avoid any mistakes they make in their use of the equipment and the stage.

Just prior to the talk, it may help to take a few minutes of "quiet time" (if possible), in which you focus on the main points you wish to convey and on how you will deliver the talk.

11.2. Delivery

Try to give a dynamic presentation that conveys enthusiasm for your subject. The total amount of time that you are taking from the audience (size of audience × length of talk) may amount to many hours, so you should make every effort to give an inspiring and professional presentation.

Two of the most common and easily avoided mistakes are to speak too quickly and to speak too quietly. Most people can be heard at the back of the room provided they face the audience and remember to speak more loudly than usual. Asking "Can you hear me at the back?" should not be necessary; it should be possible to judge from the sound of your voice and the faces of the audience whether they can hear. You should speak more slowly than in normal conversation: Kenny [149] suggests not more than 100 words per minute, while Calnan and Barabas [50] suggest an upper limit of 120 words per minute. Following this advice is not easy because nerves tend to cause us to speak more quickly than usual. Moreover, when in full flow of explanation we may feel we have so much to say that we must rush to fit it all in; practising beforehand should preclude this feeling by giving confidence in the timing of the talk. One of the reasons for speaking slowly, particularly in a large room, is that when we raise our voices the different types of sound scale at different rates. For example, we cannot increase the volume of the consonants f, k, p and t—just that of the vowels around them. Therefore a slower rate of delivery is needed to keep our speech understandable. Of course, it is also important to keep in mind that the audience may contain people whose first language is not English and for whom understanding spoken English presents difficulties.

Avoid um's, ah's and other sounds designed to fill the gap between one sentence and another. Pause instead. Pauses give the audience a moment to digest what you have just said and to anticipate what is coming next. They give you time to sense the mood and reaction of the audience and to gather your thoughts. Pausing while changing a slide gives the audience a chance to ask questions (if you do not want questions during the talk you can ask that they be held until the end). Note that a pause never seems as long to the audience as it does to the speaker.

For variety, instead of stating a fact try posing it as a question and then answering it: "Why doesn't it suffice to look at the eigenvalues? Because the matrix is highly nonnormal." Anticipate the audience's line of thinking with phrases such as "You may be wondering why...." The talk can be enlivened by the use of analogies, either from the technical field under discussion or from life in general.

Look at the audience as much as you can, trying to cover the whole audience. Eye contact is vital if you are to build up a rapport with the audience and maintain their interest. If you stare into the screen or at the projector you will miss the feedback given by the faces in the audience. Never turn your back on the audience, even when pointing; a sideways stance allows you to look at the screen or the audience with small movements of the head.

A tip I learned from a drama teacher is to exaggerate gestures made with the hands or arms: pointing with a fully stretched arm is more effective than pointing with a hand.

Nerves are perfectly natural and affect all speakers. They can be employed to advantage because they generate the energy necessary for an effective talk. Taking a few deep breaths before beginning to speak helps to control nerves. Provided that you have prepared properly, nerves should disappear once the talk is underway. Remember that the audience are on your side and want you to succeed, so do not be daunted by them.

Vary the pitch of your voice, mainly downwards, to maintain the listeners' interest. Speaking in a monotone is a sure way to send an audience to sleep. Watch out for the tendency for nervousness to cause the voice to rise in pitch, because of tightening of the muscles around the throat and voice box. Variations in the speed and rhythm of your speech are also worth aiming for.

Finishing on time is important. Overrunning indicates a lack of professionalism and is apt to lose you the audience's attention, as they ponder the coffee break, the next talk or lunch. At a conference, exceeding your time is particularly discourteous as it either takes away time from the next speaker or causes the carefully worked out schedule to be upset (a particular nuisance if there are parallel sessions). I recommend writing the finishing time in your notes before you begin. It is easy to forget, and the simple mental calculation of when you should stop can be difficult when you are speaking.

Time seems to pass more quickly for the speaker than for the audience, which is one of the reasons why some speakers overrun. To keep yourself aware of the time it is a good idea to remove your watch and place it on the table next to the slides. You can then regularly glance at the watch without the audience noticing.

Signal that the end is coming to awaken the interest of the audience and to increase the chance of your final message be remembered. If the end of the talk is well constructed it should be obvious when you have spoken your last sentence, but you may want to add "thank you" or "thank you for your attention", particularly if the chairperson seems to be dozing! Let the chairperson ask if there are any questions and select the questioners. Immediately asking for questions yourself precludes applause and takes listeners' thoughts away from your concluding sentences. In a large room, repeat a question on receiving it. This ensures that all the audience hear the question and gives you time to think of an answer. If you are not sure how to answer the question, it is perfectly acceptable to say something like "That's a very interesting question to which I don't know the answer", perhaps adding "Can we talk about it afterwards?" One word answers "Yes" or "No" are acceptable, and allow time for further questions. Do not give answers so long that they sound like a second lecture.

It is tempting to relax at the end of your talk, which can cause you to lose concentration during the question and answer session. However, you need to be fully alert to answer difficult or unexpected questions, and if you handle questions poorly you will leave the audience with a negative impression of your work.

You may like to prime one or two friends with questions, as one question is usually all it takes to break the ice and start a stream of questions.

Some more mistakes to avoid:

- Don't put a sweaty palm on the slide. It will produce a blotch on the screen that gradually fades before the audience's eyes.

- Avoid blocking the path of light from projector to the screen as well as the audience's view of the screen. I remember one speaker being

---

The Ten Commandments of Giving a Talk

1. Design the talk for the audience.

2. Prepare thoroughly and rehearse the talk.

3. Produce clear, legible slides.

4. Arrive early and check the lecture room.

5. Speak slowly and loudly.

6. Be enthusiastic about what you say.

7. Look at the audience as you speak.

8. Don't fidget with the slides or the pointer.

9. Finish on time (or early).

10. Answer questions courteously and concisely, and admit it if you don't know the answer.

Figure 11.1.

asked by a member of the audience "Can you pace nervously back and forth so that I can see the screen?"

- Avoid noisy slide management. If the slides have backing sheets, tear them off before the talk (if you don't, the act of tearing off a backing sheet does at least create a necessary pause). If you keep the slides in a ring binder, don't open and close it between each pair of slides.

- Don't leave the screen blank for more than a few seconds, as this may distract and dazzle the audience. Turn the projector off or put a piece of paper on the projector.

Figure 11.1 summarizes advice on preparing and delivering a talk.

Finally, it is important to appreciate the role that talks play in a research community. They are the focus of the community's interactions and a celebration of its achievements and they bring the subject to life. Talks provide an opportunity to gain insight into a researcher's work and personality that cannot be obtained from the printed page. Every subject has its star speakers who are renowned for the quality of their talks. You can learn a lot about your subject and about giving talks by listening to these people.

11.3. Further Reading

Excellent advice on both writing and giving a talk can be found in various places. In "How to Talk Mathematics" [123], Halmos addresses advice to the young mathematician, but what he says about such topics as simplicity, detail, organization and preparation should be read by all mathematicians (see the quotation at the beginning of the previous chapter). Forsythe [82] gives suggestions to students on how to present a talk about a mathematical paper. In *A Handbook of Public Speaking for Scientists and Engineers*, Kenny [149] discusses all aspects of public speaking for the scientist, including preparation of material, choice of visual aids, and presentation. He recommends tongue twisters for the busy scientist to improve articulation! The chapter "Speaking at Scientific Meetings" in Booth's *Communicating in Science* [36] is full of good advice. Turk's *Effective Speaking* [279] is a general guide to speaking; it reports research findings on topics such as audience psychology, memory and non-verbal communication. Calnan and Barabas [50] have written an excellent guide to *Speaking at Medical Meetings*, most of which applies to any form of technical speaking.

Byrne [48] and Garver [102] both offer practical advice on preparing and giving technical presentations. A transcript of a (presumably imaginary) talk that breaks so many rules that it is funny is given in "Next Slide Please" [10]. The American Statistical Association gave workshops on improving technical presentations at its 1980–1982 annual meetings and the material that emerged from the workshops is summarized in "Presenting Statistical Papers" [86]. Common mistakes in using an overhead projector are discussed by Gould in "The Overhead Projector" [113], and in "Visual Aids—How to Make Them Positively Legible" [114] he discusses the choice of letter size for slides, pointing out the effects on legibility of room layout and screen size and positioning. There are many general books on public speaking and much of what they say is relevant to scientific talks. One of my favourites is Aslett's *Is There a Speech Inside You?* [11].

An aspect of public speaking that is often overlooked is proper use of the voice. Two useful books in this respect are Berry's *Your Voice and How to Use It Successfully* [29] and Rodenburg's *The Right to Speak* [237], both of which contain advice on pronunciation, breathing and relaxation.

Memory systems can help you to remember your notes, or facts and figures not shown on the slides. See any of the books by Buzan or Lorayne, including [46], [183], [184].

Chapter 12
Preparing a Poster

> Poster sessions, pioneered in Europe,
> made their first appearance at a major U.S. meeting
> in Minneapolis earlier this month
> at the Biochemistry/Biophysics 1974 Meeting.
>
> — THOMAS H. MAUGH II, *Poster Sessions:
> A New Look at Scientific Meetings* (June 1974)

> *Visualizing a poster in its entirety at the planning stage is a must.*
> — JOHN D. WOOLSEY, *Combating Poster Fatigue* (1989)

> *Calvin: Want to help me make a poster?...*
> *Hobbes: Sounds good. What's our winning poster going to say?*
> *Calvin: That's where you come in.*
> — CALVIN, *Calvin and Hobbes by Bill Watterson* (1994)

> *A picture is worth a thousand words.*
> — Anonymous

179

Poster sessions are an increasingly important part of scientific conferences. In this chapter[19] I offer some advice on the preparation and presentation of posters, something in which many of us are rather inexperienced.

12.1. What Is a Poster?

A poster is very different from a paper or a talk, and so different techniques need to be used in its preparation. In particular, a poster is not a conference paper, and simply pinning a paper to a poster board usually makes a very poor poster. A poster board is typically 4 feet high and 6 feet wide, but the reverse orientation (tall and thin) is also seen. It is advisable to check beforehand on the size of the boards that will be available to you. A poster itself is a visual presentation comprising whatever the contributor wishes to display on the poster board. Usually, a poster is made up entirely of sheets of paper pinned or attached with velcro strips to the board, but there is no reason why other visual aids should not be used. The pins or velcro are usually provided with the board by the conference organizers.

The purpose of a poster is to outline a piece of work in a form that is easily assimilated and stimulates interest and discussion. The ultimate aim is a fruitful exchange of ideas between the presenter and the people reading the poster, but you should not be disappointed if readers do not stop to chat—a properly prepared poster will at least have given useful information and food for thought.

12.2. A Poster Tells a Story

In preparing a poster, simplicity is the key. A typical reader may spend only a few minutes looking at the poster, so there should be a minimum of clutter and a maximum of pithy, informative statements and attractive, enlightening graphics. A poster should tell a story. As always in a scientific presentation, the broad outline includes a statement of the problem, a description of the method of attack, a presentation of results, and then a summary of the work. But within that format, there is much scope for ingenuity. A question-and-answer format, for example, may be appropriate for part of the poster.

A poster need not have the same title as a paper on which it based— a more declarative or provocative title may help to grab attention. Thus "Accuracy and Stability of the Null Space Method for Solving the Equality Constrained Least Squares Problem" could be replaced by "The Null Space Method is Stable". Short section titles such as "Introduction", "Analysis",

---

[19]This chapter is based on [129].

"Experiments" may be less effective at drawing the reader in and providing an overview of the poster than longer, more informative phrases. For example, the first section might be titled "Three variations of the null space method" (omitting a tempting initial "There are"), and the section reporting experiments might be headed "The bound provides good error estimates in practice."

A poster should not contain a lot of details—the presenter can always communicate the fine points to interested participants. In particular, it is not a good idea to present proofs, except in brief outline, unless the proofs are the focus of the presentation. Keep in mind that the poster will be one of many in the exhibition area: you need to make sure that it will capture and hold the reader's attention.

The poster should begin with a definition of the problem, together with a concise statement of the motivation for the work. As when producing slides for a talk, it is not necessary to write in complete sentences; sentence fragments may be easier to comprehend. Bulleted lists are effective. An alternative is to break the text into chunks—small units that are not necessarily paragraphs in the usual sense.

An abstract should not be necessary, since it may already appear in the conference program and it would (presumably) repeat what is already concisely described on the poster.

For presenting results, graphs—easier to scan than the rows and columns of figures in a table—are even more appropriate than in a paper, but they should be simplified as much as possible. In particular, labels should be minimal. A brief description of the implications of a graphic, placed just above or below it, is helpful. For ideas on graphic design, a wide selection of books is available, such as any of those by Tufte [275], [276], [277].

Conclusions should be brief and they should leave the reader with a clear message to take away.

12.3. Designing Your Poster

A poster is usually formed from separate sheets of letter paper: 8×11 inches (US) or A4 (Europe). The number of pages should be minimized; for these sizes a suggested maximum is 15. But larger sheets, or even sheets of differing sizes within one poster, can also be effective.

It is worth repeating that it is definitely unacceptable to post a copy of a paper.

The typeface should be considerably larger than for a normal document. Not all readers will have perfect eyesight and light levels may be low. Moreover, the crowd of readers around a popular poster may be several people deep, so the type must be easily readable by a person standing a few feet away. The title of the poster and the author's name should be large and prominent, but should not overwhelm the poster.

If it is not convenient to print directly at the desired typesize, pages can be magnified on a photocopier. Good use can be made of colour, both to provide a more interesting image and for colour coding of the text. A coloured backing card for each sheet can be effective. For added interest, try including an appropriate cartoon, photograph, or quotation. There is plenty of scope for creativity.

The text itself should be composed of short sentences and paragraphs. To transmit the important information effectively, try to put the key terms near the start of each sentence. Thus, instead of writing

> In designing or choosing a summation method to achieve high accuracy, the aim should be to minimize the absolute values of the intermediate sums T_i.

write

> For high accuracy, minimize the absolute values of the intermediate sums T_i.

(That this statement applies to summation is assumed to be clear from context.)

If the sheets are arranged as a matrix, two layouts are possible: horizontal (reading across the rows) and vertical (reading down the columns). While the horizontal ordering is perhaps more natural, it has the major disadvantage of requiring the reader to move to and fro along the poster; if there are many readers, congestion can result. A vertical ordering is therefore preferable, although other possibilities should be considered as well. If you are comparing three methods, for example, you could display them in parallel form, in three rows or columns, perhaps as a "display within a display."

Consider the possibility of arranging the poster to represent some feature of the problem, such as a particular sparsity structure of a matrix. If there is any doubt about the order in which the sheets should be read, guide the reader by numbering the sheets clearly or linking them with arrows.

Think carefully about the use of the poster board. One extreme is to spread the sheets out to make full use of the board—taking care to position them at a height at which they can be read by both the short and the tall. If there are only a few sheets, it may be best to concentrate them in a small area, where a reader can proceed from beginning to end while standing in one position. Several examples of layout and further discussion are given by Matthews [195]. Woolsey [301] offers many useful suggestions on poster design and gives two informative annotated examples of posters describing his paper—one good, one bad. Other useful references are Briscoe's *Preparing Scientific Illustrations* [38] and Anholt's *Dazzle 'em with Style* [9]

12.4. Transportation and the Poster Session

Transporting a poster can be a problem if it contains large sheets of paper. Rolling the paper into a cylinder is the most common system. You will usually be allotted plenty of time to set up the poster, so it may be easiest to bring it in pieces, to be assembled on site (but be sure to work out the layout beforehand, and bring a diagram). If the work presented in the poster has been described in more detail in a paper, consider making the paper available as a handout at the poster session.

Once the session starts, stand near the poster but not in a position that obscures it from view. Be prepared to answer the questions that a good poster will inevitably generate. But keep in mind the advice of one expert: "A presenting author at a poster session should behave like a waiter in a first-class restaurant, who is there when needed but does not aggravate the guests by interrupting conversation every ten minutes to inquire whether they are enjoying the food" [9].

12.5. A Word to Conference Organizers

If we wish presenters to take poster sessions seriously, and if we want the submission of a poster to be seen as a viable alternative to giving a talk, then it behoves conferences organizers not to treat the presenters as second-class citizens. This means making poster sessions an integral part of the conference program, providing appropriate facilities for the setting up and presentation of posters, and encouraging conference participants to attend the poster sessions. Proper time should be allowed in the program for the poster sessions; adequate boards, fasteners, and space should be provided, and the poster rooms should not be remote from the rest of the conference. If the dining area is large enough, consider having some posters there—a good audience is assured! A poster prize is also worth considering.

Chapter 13
T_EX and L^AT_EX

*T_EX can print virtually any mathematical thought that comes into your head,
and print it beautifully.*

— HERBERT S. WILF, *T_EX: A Non-review* (1986)

*I can't compose at a typewriter.
I can't even compose a letter to my relatives on a typewriter,
even though I'm a good typist.*

— DONALD E. KNUTH, in *Mathematical People:
Profiles and Interviews* (1985)

*1992: Extensive testing shows that 98.3% of the time
no matter which of the* [h], [t], [b], *or* [p] *options is used,
L^AT_EX will put your* table *at the end of the document.*

— DAVID F. GRIFFITHS and DESMOND J. HIGHAM,
Great Moments in L^AT_EX History[20] (1997)

*If you don't find it in the index,
look very carefully through the entire catalogue.*

— SEARS, ROEBUCK and CO., *Consumer's Guide* (1897)

---

[20]In [118].

Wordprocessing systems have brought a new dimension to writing. Courses
are offered on writing with a computer, and books have been written on
the subject [205], [302]. Producing mathematical equations is difficult or
impossible with wordprocessors not designed for the purpose, but various
specially designed systems are available. Because it is ubiquitous in the
mathematical and computer science communities I focus in this chapter
on the TₑX typesetting system (the "ₑX" is usually pronounced like the
eck in *deck*) and the LⁱTₑX macro package. The `troff` system (with the
`eqn` preprocessor for typesetting mathematics [151]) predates TₑX and was
widely used on Unix systems, but has now largely been supplanted by TₑX.
My purpose is not to describe or teach TₑX and LⁱTₑX or the associated
programs BɪʙTₑX and MakeIndex (see the references in §13.5 for suggested
texts) but to explain their advantages and to give hints on their use.

13.1. What are TₑX and LⁱTₑX?

TₑX was developed by Donald Knuth at Stanford University, beginning in
the late 1970s [161]. Knuth originally designed TₑX to help him typeset
books in his *Art of Computer Programming* series. It is a program that
accepts a text file interspersed with formatting commands and translates it
into a `dvi` (device independent) file that can readily be translated into the
languages understood by output devices such as screens and laser printers.
The typical usage is to prepare an ASCII file using your favourite editor,
run it through TₑX and preview the output on the screen. You correct
errors by editing the source file and then re-running TₑX. When you wish
to obtain hard copy, you invoke a driver to render the TₑX output on the
printer.

LⁱTₑX [172] by Leslie Lamport is a macro package that "sits on top" of
TₑX. It was originally released in 1985 and rapidly became the package of
choice for TₑX-using authors in computer science, mathematics, and related
areas. I recommend using LⁱTₑX rather than TₑX, since it provides so many
useful features that would have to be programmed if you used TₑX. The
original version of LⁱTₑX, LⁱTₑX 2.09, was extended in various ways by many
people as its popularity grew, and so suffered from incompatible dialects.
This problem was solved by LⁱTₑX 2ₑ, which in 1994 replaced LⁱTₑX 2.09 as
the official version of LⁱTₑX. It is supported and maintained (with a new
release every six months) by volunteers in the LⁱTₑX3 Project, a long term
project to produce a completely re-designed and re-implemented version of
LⁱTₑX. In the remainder of this chapter, "LⁱTₑX" refers to LⁱTₑX 2ₑ.

Another macro package for TₑX called 𝒜ₘ𝒮-TₑX [251], by Michael Spivak, was designed specifically for typesetting mathematics. It introduced,
among other features, special fonts, flexible structures for multiline dis-

played equations, and facilities for producing commutative diagrams. It was developed for use by the publications of the American Mathematical Society (AMS). $\mathcal{AMS}$-LATEX from the AMS [8] provides a package `amstex` for LATEX that incorporates the features of $\mathcal{AMS}$-TEX.

There are several advantages to preparing your papers in some form of TEX.

1. TEX is available on many computer systems, from PCs to workstations to mainframes, and output can be obtained on devices ranging from screens to dot matrix and laser printers to professional typesetting machines.

2. Many journals now accept papers in TEX format; this saves the usual typesetting stage and its inevitable introduction of errors.

3. In LATEX and $\mathcal{AMS}$-TEX the format of the document is governed by style options, which are specified at the beginning of the source. Only the style options need to be changed to alter the format of the document. Available formats range from book to technical report to SIAM or AMS journal paper.

4. TEX source is plain ASCII text so it can readily be sent by electronic mail or ftp'ed. This is convenient when collaborating with a distant co-author or submitting a final paper in TEX format to a journal.

5. TEX's `dvi` files, or their PostScript translations, are relatively portable means for distributing papers electronically. For example, you can make your papers available from a Web page or by anonymous ftp (see §14.1). (Making the `tex` file available is less satisfactory because it is often inconvenient to collect all the source into one file, particularly if LATEX is used with nonstandard style files.)

6. Cross-references to equations, theorems, tables, etc. and citations of bibliographic references are fully supported using symbolic labels in LATEX. Renumbering of the cross-references is handled automatically when a new equation or bibliographic entry is added.

7. It is easy to prepare transparencies with TEX if the paper on which the talk is based is already in TEX form, as the conversion largely consists of deleting chunks of text and equations. See §10.3.

13.2. Tips for Using LATEX

In this section I offer some suggestions on the use of LATEX, including some finer points not often discussed in textbooks (many of the suggestions apply to TEX and $\mathcal{A}_{\mathcal{M}}\mathcal{S}$-TEX too). Appendix B summarizes the TEX and LATEX symbols and spacing commands.

Dashes

There are three types of dashes in typesetting. A hyphen is produced in TEX by typing -, as in state-of-the-art computer. Number ranges use the en-dash, typed in TEX as -- rather than a single dash. This gives 1975–1978 and pp. 12–35 instead of 1975-1978 and pp. 12-35. The en-dash is also used when the first part of a compound word does not modify the meaning of the second part; it can be thought of as standing for *and* or *to*. Examples: Gauss–Chebyshev quadrature, Moore–Penrose inverse, left–right evaluation. An exception is Runge-Kutta method, for which most authors use a hyphen. A dash partitioning a sentence—like this—should be typed as --- with no space on either side; this is the em-dash. A mathematical minus sign is produced by a single - within mathematics mode. Compare -3, typed as -3, with -3, typed as -3.

Delimiters

Be wary of the \left and \right commands for automatically sizing delimiters, as they often produce symbols that are too large (sometimes depending on taste). One of the \big, \Big or \bigg commands (in order of increasing size) usually gives the appropriate size. Here are three examples.

$$c(A) = \min\left\{ \epsilon : \sum_{i=0}^{\infty} |G^i M^{-1}| \leq \epsilon |(I - G)^D M^{-1}| \right\}, \quad (\text{\left}),$$

$$c(A) = \min\biggl\{ \epsilon : \sum_{i=0}^{\infty} |G^i M^{-1}| \leq \epsilon |(I - G)^D M^{-1}| \biggr\}, \quad (\text{\bigg}),$$

$$S_n = \sqrt{\frac{2}{n+1}} \left(\sin\left(\frac{ij\pi}{n+1} \right) \right)_{i,j=1}^{n}, \quad (\text{\left}),$$

$$S_n = \sqrt{\frac{2}{n+1}} \Bigl(\sin\bigl(\frac{ij\pi}{n+1} \bigr) \Bigr)_{i,j=1}^{n}, \quad (\text{\Big inner, \bigg outer}),$$

$$G\left(x^{(k)} \right) = -g\left(x^{(k)} \right), \quad (\text{\left}) \qquad G\bigl(x^{(k)} \bigr) = -g\bigl(x^{(k)} \bigr) \quad (\text{\big}).$$

Figures in LaTeX

Figures can be drawn in LaTeX using the `picture` environment, which has a repertoire of circles, ovals, and straight lines at a limited number (48) of slopes. (Figure 13.1, below, was drawn this way.) Because of the limitations of TeX for drawing pictures it is often necessary to produce figures with a program designed specifically for the purpose. This is most commonly done using figures represented in PostScript.[21]

The standard way to include PostScript figures in LaTeX is via the `\includegraphics` command in the `graphics` or `graphicx` packages (the latter offers more options and a key–value interface). These packages replace the earlier `epsf` and `psfig` style files that were commonly used with LaTeX 2.09. The usage is illustrated by

```
% Assumes the use of the dvips dvi to PostScript driver.
\usepackage[dvips]{graphicx}
...
\begin{center}
% Include eps file, applying scale factor 0.6.
\includegraphics[scale=0.6]{figure.eps}
\end{center}
```

For much more on generating and manipulating PostScript files with LaTeX, see the references in §13.5.

File Names and Internet Addresses

File names, email addresses and URLs of objects on the Web can cause formatting problems when typed within a paragraph because TeX may not be able to find a suitable line break within them. A solution is to use the LaTeX `path` package, available from the CTAN archives (see §13.5). It provides a command `\path` that is similar to the command `\verb` but which allows hyphen-less line breaks at punctuation characters. For example, I typed the anonymous ftp locator `ftp://ftp.ma.man.ac.uk/pub/narep/narep305.ps.gz` as

```
\path"ftp://ftp.ma.man.ac.uk/pub/narep/narep305.ps.gz"
```

---

[21] PostScript is a page description language of Adobe Systems, Inc. A PostScript file is an ASCII file of PostScript language commands that specifies how the output is to appear on the page. By convention, the first line of a PostScript file begins with the two characters `%!`. Ghostscript is an interpreter for PostScript that is useful for printing PostScript files on non-PostScript printers and previewing them on the screen. It was written by Aladdin Enterprises, Palo Alto, and runs on various Unix machines and under DOS and Windows.

Labels

It is worth using a consistent and numbering-independent system for the choice of keys in \label commands. For example, rather than using the keys for equations eqn1, eqn2, etc., try g.def (definition of function g), CS-ineq (Cauchy–Schwartz inequality) and bound1 (first of several bounds). These more meaningful keys are easier to remember and make the source code easier to navigate. For figures, tables, theorems, lemmas and so on I prefer to use keys of the form x.y, where x is an abbreviation for the type of object: Fig.residuals, Lem.limit, Thm.Fermat, etc.

The showlabels package, available from the CTAN archives, can be useful at the drafting stage: it causes the key to be shown in the margin at each instance of a \label command.

Macros

It is a good policy to use macros to typeset important notation, whether it be words or symbols. As well as reducing the amount of typing and ensuring consistency, this helps to emphasize logical structure in the source, which will be appreciated if you have to change the notation, for only the macro definitions will need to be altered. The latter point is discussed by Lamport [171], who argues against WYSIWYG (what you see is what you get) systems and in favour of markup systems such as LATEX. As an example, among the macros used in typesetting this book are[22]

```
\def\mw{mathematical writing}
\def\norm#1{\|#1\|}
\def\Htk{\widetilde{H_k}}
```

The first macro is used to save typing: I type \mw instead of "mathematical writing". I find the second macro makes the source code easier to read and modify: I type \norm{x} to produce $\|x\|$. It makes it easy to change from a general norm to a specific one if the need arises (e.g., to convert to the 2-norm the definition is changed to \def\norm#1{\|#1\|_2}). The third macro is used in the source for §3.5; again, it saves typing and simplifies any future change of notation.

Miscellaneous Mathematics

The colon character is treated as a relational operator in mathematics mode, as in $x := y$ and $\{z : |z| \leq 1\}$. If you want a colon to be

---

[22]The LATEX user is recommended to define macros with the \newcommand command rather than \def. With \newcommand an attempt to define an existing command produces an informative error message, while if \def is used, an error can result that is hard to track down. I tend to use \def anyway!

Table 13.1. Confusable pairs of symbols.

Relation or binary operation symbol	Example	Ordinary symbol	Example
\mid	$\{\,x \mid x > 0\,\}$	\vert or |	$|z|$
\setminus	$G \setminus H$	\backslash	$p \backslash n$
\parallel	$\vec{u} \parallel \vec{v}$	\Vert or \\|	$\|A\|$
\perp	$\vec{u} \perp \vec{v}$	\bot	$x_\perp$
\in	$x \in \mathbb{R}$	\epsilon	$\epsilon > 0$

treated as a punctuation mark use \colon. Thus, $A(1{:}r, 1{:}r)$ is typed as $A(1\colon r,1\colon r)$. Table 13.1 lists some pairs of symbols that might be confused.

To express that one quantity is much bigger or much less than another, use the \gg or \ll symbol.

Bad: $x >> y$ (typed as $x >> y$).
Good: $x \gg y$ (typed as $x \gg y$).

Similarly, to obtain a norm symbol $\|$ type \\| rather than ||.

Recall that mathematical functions should be typeset in roman (see page 32). This rule is frequently violated by TEX users, giving aesthetically disturbing results such as $sign(det(A)) = -1$ instead of $\mathrm{sign}(\det(A)) = -1$. If you are using a function for which there is not a predefined macro, define one of the form

```
\def\diag{\mathop{\mathrm{diag}}}
```

Then type formulas such as $D = \mathrm{diag}(a_{ii})$ as $D = \diag(a_{ii})$. More generally, any term with more than one letter should normally be set in roman, such as tol, representing a tolerance. An exception is in program or algorithm listings, where, for example, italics might be used for variables.

I prefer to use the accents \widehat and \widetilde ($\widehat{x}$, $\widetilde{x}$) instead of \hat and \tilde ($\hat{x}$, $\tilde{x}$) because the wide versions are easier to read and are less likely to be mistaken for a stray blob on a photocopied page.

Exercise restraint in the use of TEX's mathematical formatting within the text—elaborate formulas can look out of place and tall ones can affect the line spacing. A common mistake is to use the LATEX command \frac instead of a slash: $x = (y + z)/2$ looks better than $x = \frac{y+z}{2}$, and $\frac{a}{b}$ is better written as ab^{-1} or a/b.

An ellipsis (the sequence of three dots "...") is used to mark omission from text or in a mathematical formula such as "$i = 1, 2, \ldots, n$." In both

cases if you type the ellipsis as three full stops in TEX the spacing is too
tight: $i = 1, 2, ..., n$. Instead, the control sequence \dots should be used (or
\cdots for centred dots in a formula—see page 32). Vertical and diagonal
ellipses for use in matrices are obtained with

\vdots (⋮) and \ddots (⋱).

Compare the two expressions

$$S = A_{22} - A_{21}A_{11}^{-1}A_{21}^T,$$
$$S = A_{22} - A_{21}A_{11}^{-1}A_{21}^T.$$

Notice that in the first the subscripts in the third term are not on the
same horizontal level. This has been corrected in the second expression by
typing A_{21}^{}, where the empty superscript has the effect of lowering
the subscript to the desired level. Whether you think this sort of refinement
is worth the trouble depends on your aesthetic judgement.

Which of the following two displays do you prefer?

$$\Pi^T(A + \Delta A)\Pi = (R + \Delta R)^T(R + \Delta R),$$
$$\Pi^T(A + \Delta A)\Pi = (R + \Delta R)^T(R + \Delta R).$$

The second differs from the first only in that the Greek letters have been
made italic by the use of the \mathit command. I prefer to put uppercase
Greek letters in the italic font because they then mesh better with the italic
roman letters, particularly in expressions such as ΔA.

Finally, as an example of how TEX's spacing can sometimes be fine-
tuned, compare

$\Pi^T A\Pi$, typed as $\mathit{\Pi}^T A\mathit{\Pi}$,

$\Pi^T A\Pi$, typed as $\mathit{\Pi}^T\mskip-5mu A\mathit{\Pi}$.

The small negative space \mskip-5mu, where the value -5 was obtained
by experimentation, removes the unattractive extra space between the Π^T
and the A.

Quotes, Dates, Lists and Paragraphs

Be careful to distinguish between the opening quote (' or \lq) and the
closing quote (' or \rq). Errors such as "unmatched quotes" are irritating
to the trained eye. Do not use the double quote ", which always produces
the closing quotes ".

In LATEX set the date explicitly using the \date command (assuming
you use \maketitle)—otherwise the date that is printed will be the date
on which the file was LATEXed, which is less informative.

Table 13.2. Order of invocation of LaTeX, BibTeX and MakeIndex.

LaTeX only	LaTeX and BibTeX	LaTeX, BibTeX and MakeIndex
latex	latex	latex
latex	bibtex	bibtex
	latex	latex
	latex	latex
		makeindex
		latex

Be careful not to overuse lists. LaTeX makes it very easy to produce bulleted or numbered lists, with the `itemize` and `enumerate` environments. An inappropriate use of lists is to present a series of ideas without linking them smoothly. A list may be the remnant of an outline that has not been fully developed. If a list is longer than a full page, or has more than one paragraph in an entry, you should consider converting it to running text.

It is easy to start new paragraphs unintentionally in TeX by leaving a blank line after a list, quotation or mathematical display. Watch out for this error, as it is easy to overlook when you are revising and proofreading and it can be disconcerting for the reader.

Running LaTeX, BibTeX and MakeIndex

To ensure that cross-references, references to the bibliography, and index entries are correctly resolved, you need to run LaTeX more than once. Table 13.2 shows the necessary sequences of invocations. However, in certain circumstances more invocations will be required. For example, if a BibTeX entry is used that itself contains a `\cite` command (most likely in a `note` field) then BibTeX needs to be run twice:

latex, bibtex, latex, bibtex, latex, latex

Always check the `log` file for warning messages before printing a `dvi` file.

Source Code

When typing your paper make liberal use of spaces and indentation to improve the readability of the source. Start new sentences and main phrases on new lines. This approach simplifies editing: the source is easier to navigate, and cut and paste operations are easier because they tend to act on whole sentences or phrases. It may also be desirable to limit lines to

72 characters, for reasons explained in §8.7. Here is a short extract from a paper of mine (in LATEX) that illustrates these points:

```
We assume that $x_{k+1}$ is computed by
forming $Nx_k+b$ and then solving a linear system with $M$.
The computed vectors $\xhat_k$ therefore satisfy
\begin{equation}
        (M + \dM_{k+1})\xhat_{k+1} = N\xhat_k + b + f_k,
   \label{be1}
\end{equation}
```

Spacing in Formulas

Correct spacing in displayed formulas can usually be accomplished with the \quad and \qquad commands. A \quad is a printer's quad of horizontal space, which is approximately the width of a capital M, and a \qquad is twice the width of a \quad. A \qquad is appropriate when a formula is followed by a qualifying expression. Less often, a thick space, typed as \; (5/18 of a quad), is useful. Here are some examples of spacing.

$$\|A^T\|_p = \|A\|_q, \qquad \frac{1}{p} + \frac{1}{q} = 1,$$

$$fl(x \text{ op } y) = (x \text{ op } y)(1 + \delta), \qquad |\delta| \le u, \quad \text{op} = *, /,$$

$$\alpha_i = \alpha_j \quad (i < j) \quad \Rightarrow \quad \alpha_i = \alpha_{i+1} = \cdots = \alpha_j.$$

$$\eta(y) = \min\{ \epsilon : (A + E)y = b + f, \ \|E\| \le \epsilon\|A\|, \ \|f\| \le \epsilon\|b\| \}.$$

These displays were typed in LATEX as follows (I have wrapped some of the lines so that they fit comfortably in the given page width).

```
\def\op{\mathbin{\mathrm{op}}}   % Binary operator
\def\norm#1{\|#1\|}
\begin{eqnarray*}
    \norm{A^T}_p &=& \norm{A}_q, \qquad
                    \frac{1}{p} + \frac{1}{q} = 1, \\
    fl(x \op y) &=& (x \op y)(1+\delta),
                    \qquad |\delta|\le u, \quad \op = *,/,
\end{eqnarray*}
$$
    \alpha_i = \alpha_j \quad (i<j) \quad \Rightarrow
    \quad \alpha_i = \alpha_{i+1} = \cdots = \alpha_j.
$$
$$
```

```
\eta(y) = \min \{\, \epsilon: (A+E)y = b+f, \;\;
                 \norm{E} \le \epsilon \norm{A},
                 \;\; \norm{f} \le \epsilon \norm{b} \,\}.
$$
```

Note the thin spaces (obtained with \, and of width 1/6 of a quad) placed inside the parentheses in the definition of η, which are recommended for sets defined by a generic element followed by a specific condition [161, p. 174]. Note also that the TeX construct $$..$$ is more properly typed in LaTeX as \[..\] or as a `displaymath` environment.

A \quad is also useful for separating words from symbols:

$$\text{minimize} \quad F(x) \quad \text{subject to} \quad c(x) = 0,$$
$$y'(t) = F(t, y(t), y(t - \tau)), \quad y(t) = \Phi(t) \text{ for } t \in [-\tau, 0],$$
$$g(n) \ge \frac{n+1}{2} \quad \text{for all } n.$$

These displays were typed in LaTeX as follows.

```
\begin{eqnarray*}
  & \mbox{minimize} \quad F(x) \quad \mbox{subject to}
                         \quad c(x) = 0 & \\
  &  y'(t) = F(t, y(t), y(t-\tau)),
                     \quad \mbox{$y(t) = \Phi(t)$
                         for $t\in[-\tau,0]$} & \\
  & g(n) \ge \displaystyle\frac{n+1}{2}
                     \quad \mbox{for all $n$}. &
\end{eqnarray*}
```

When breaking a displayed formula before a plus or minus sign (see §3.7), you need to type {}+ or {}- so that TeX knows the sign is acting as a binary operator. Without the {}, the lesser spacing assigned to a unary operator is produced. Compare these two `eqnarrays`, where the second line of the second is typed as && {}+ y + z:

$$A = v + w + x \qquad\qquad A = v + w + x$$
$$+y + z \qquad\qquad\qquad + y + z$$

By default, LaTeX puts rather too much space around the centre column of an `eqnarray` environment. A simple cure, used in this book, is to use the `eqnarray` package of David M. Jones, located at `ftp://theory.lcs.mit.edu/pub/tex/eqnarray.sty`

Ties and Spaces

Avoid unfortunate line-breaks by binding units together with a tie (~). For example, as a matter of course type `Theorem~3` not `Theorem 3`, `Mr.~Sandman` not `Mr. Sandman`, and `1,~2 and~3` not `1, 2 and 3`. Other bad breaks, such as between the last two words in a paragraph, can be corrected as and when they appear.

Leave a space before the `\cite` command.

    Bad: `Knuth\cite{knut86}`, giving Knuth[161].
    Good: `Knuth \cite{knut86}`, giving Knuth [161].
    Good: `Knuth~\cite{knut86}`, giving Knuth [161].

The "Knuth[161]" form is a common mistake. The last form avoids a line break just before the citation.

TEX assumes that a full stop ends a sentence unless it follows a capital letter. Therefore you must put a control space after a full stop if the full stop does not mark the end of the sentence. Thus p. 12 is typed as `p.\ 12`, and cf. Smith (1988) as `cf.\ Smith (1988)`. A less obvious example: MATLAB (The MathWorks, Inc.) is typed as

`\textsc{Matlab} (The MathWorks, Inc.)\`

since otherwise TEX will put extra space after the right parenthesis. In the references in a `thebibliography` environment there is no need to use control spaces because this environment redefines the full stop so that it does not give end of sentence spacing.

On the other hand, if a sentence-ending full stop follows an uppercase letter you must specify that the full stop ends the sentence. In LATEX this can be done by inserting the `\@` command before the full stop, as in the sentence from page 11

`There is also an appendix on how to prepare a CV\@.`

The `\@` command is needed more often than you might realise—in this book there are over 10 occurrences of the command.

13.3. BIBTEX

BIBTEX, written by Oren Patashnik, is a valuable aid to preparing reference lists with LATEX. To use BIBTEX you need first to find or create a `bib` file that comprises a database of papers containing those you wish to cite. BIBTEX reads a LATEX `aux` (auxiliary) file and constructs a sorted reference list in a `bbl` file, making use of the `bib` file. This reference list is read and

processed by LaTeX. A diagram showing how LaTeX interacts with BIBTEX
and MakeIndex is given in Figure 13.1 (MakeIndex is described in the next
section). In this section I go into some detail on the use of BIBTEX because
it tends not to be emphasized by the books on LaTeX and yet is a tool that
can benefit every serious LaTeX user.

A `bib` file is an ASCII file maintained by the user (in the same way as
a `tex` file). As an example, references [158] and [161] are expressed in the
BIBTEX file used for this book as[23]

```
@article{knut79,
  author = "Donald E. Knuth",
  title = "Mathematical Typography",
  journal = "Bull. Amer. Math. Soc. (New Series)",
  volume = 1,
  number = 2,
  pages = "337-372",
  year = 1979
    }


@book{knut86,
  author = "Donald E. Knuth",
  title = "The {\TeX book}",
  publisher = "Addison-Wesley",
  address = "Reading, Massachusetts",
  year = 1986,
  pages = "ix+483",
  isbn = "0-201-13448-9"
    }
```

The first part of the last sentence was typed as

```
As an example, references \cite{knut79} and \cite{knut86} are
expressed in \BibTeX\ format as
```

Once you have built up a few entries, creating new ones is quick, because
you can copy and modifying existing ones.

There are three advantages to using BIBTEX.

1. By specifying the appropriate bibliography style (`bst` file) from the
   many available it is trivial to alter the way in which the references
   are formatted. Possibilities include BIBTEX's standard **plain**, **abbrv**
   and **alpha** formats, illustrated, respectively, by

---

[23]My `bib` file uses abbreviations (described below) for the journal, publisher and address fields, but in this example I give the fields explicitly, for simplicity.

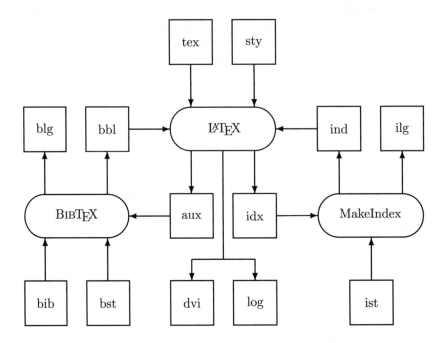

Figure 13.1. Interaction between LATEX, BIBTEX and MakeIndex. The log, blg and ilg files contain error messages and statistics summarizing the run. The aux file contains cross-referencing information. The sty, bst and ist files determine the styles in which the document, reference list and index are produced.

[1] Donald E. Knuth. Mathematical typography. *Bulletin Amer. Math. Soc. (New Series)*, 1(2):337–372, 1979.

[1] D. E. Knuth. Mathematical typography. *Bulletin Amer. Math. Soc. (New Series)*, 1(2):337–372, 1979.

[Knu79] Donald E. Knuth. Mathematical typography. *Bulletin Amer. Math. Soc. (New Series)*, 1(2):337–372, 1979.

With the alpha format, citations in the text use the alphanumeric label constructed by BIBTEX ([Knu79] in this example). The unsrt format is the same as plain except it lists the entries in order of first citation instead of alphabetical order (as is required, for example, by the journal *Computers and Mathematics with Applications*). LATEX

packages and corresponding BIBTEX style files are also available that
produce citations and bibliographies conforming to the Harvard system (see §6.11).

In this book I am using my own modification of the `is-plain` style
by Nelson Beebe, which itself is a modification of the `plain` style to
add support for ISSN and ISBN fields and for formatting a pages field
for books.

2. To keep the reference lists of working papers up to date (as technical
   reports become journal papers, for example) it is necessary only to
   update the master `bib` file and rerun LATEX and BIBTEX on each
   paper. If BIBTEX were not used, the reference list in each paper
   would have to be updated manually. Using BIBTEX also saves on
   storage, for the `bbl` files can be deleted once typesetting is complete
   and reconstructed when required by running BIBTEX.

3. Since BIBTEX is widely used and available on a wide range of machines
   it is possible for people to exchange and share databases. Two large
   collections of `bib` files deserve particular note.

   (a) BibNet is maintained by Stefano Foresti, Nelson Beebe and Eric
       Grosse. It has the URL `ftp://ftp.math.utah.edu/pub/bibnet`,
       and is also available from netlib (see §14.1), specifically from

       `http://netlib.bell-labs.com/netlib/bibnetfaq.html`

       The `bib` files in BibNet include ones containing all the publica-
       tions by particular authors (e.g., Gene Golub), and all (or many
       of) the publications that have appeared in particular journals
       (e.g., *SIAM Review*).

   (b) The Collection of Computer Science Bibliographies (which in-
       cludes bibliographies on mathematics), is maintained by Alf-
       Christian Achilles at

       `http://liinwww.ira.uka.de/bibliography/index.html`

       This large collection of bibliographies (which includes all those
       in BibNet) has an excellent Web interface and powerful search
       facilities.

   It is, of course, advisable to check the accuracy of any entries that
   you have not created yourself, before using them.

Aside from these BIBTEX-specific reasons, there are other reasons for
keeping a personalized computer database of references. If you record every

paper you read then a computer search allows you to check whether or not you have read a given paper. This is a very handy capability to have, as once you have read more than a couple of hundred papers it becomes difficult to remember their titles and authors. Also, you can put comments in the database to summarize your thoughts about papers. Some BIBTEX users include an `annotate` field in their `bib` files. Although standard BIBTEX style files ignore this field, other style files are available that reproduce it, and they can be used to prepare annotated bibliographies. In my BIBTEX book entries I include an `isbn` field.

Here are some tips on using BIBTEX.

1. To make it easier to navigate a `bib` file with your editor, keep the entries in alphabetical order by author, and use positioning lines of the form

   ```
   -md- -me-
   ```

   to indicate the start of a group of authors—this example marks the start of the authors whose last names begin with "Me", there being no last names beginning "Md".

2. Although it is tempting to save time by abbreviating `bib` file entries— typing initials instead of full author names (when they are given in the original reference), and omitting journal part numbers or institution addresses for technical reports—my experience is that it is false economy, because these details are often required eventually: for example, by a copy editor who queries an incomplete reference on page proofs. It is worth spending the extra time to create fully comprehensive entries. Note that if names are typed with no space between the initials (e.g., `author = D.E. Knuth`), BIBTEX produces only the first initial. You should therefore always leave a space between initials (`D. E. Knuth`), as is standard practice in typesetting.

3. Various conventions are in use for choosing keys for `bib` file entries (they appear on the first line of the entry and are specified in the `\cite` command). My method is to use the first four letters of the author's last name followed by the last two digits of the year. If there are two authors I use the first two letters of each author's last name and if three or more, the first letter of the first three or four authors' last names. Multiple papers in the same year by the same authors are distinguished by extra letters 'a', 'b' and so on. This method has proved effective for `bib` files with a few thousand entries. Note that a key does not have to be of the form `author:gnus`, as some people

have presumed on reading Lamport's whimsical examples involving gnus [172]!

4. BibTeX allows the use of abbreviations. For example, instead of typing

```
publisher = "Society for Industrial and
             Applied Mathematics",
address = "Philadelphia",
```

you can type

```
publisher = pub-SIAM,
address = pub-SIAM:adr,
```

where the following string definitions appear prior to these lines in the `bib` file (typically at the start of the file):

```
@STRING{pub-SIAM  = "Society for Industrial and
                     Applied Mathematics"}
@STRING{pub-SIAM:adr = "Philadelphia"}
```

The use of abbreviations saves typing and ensures consistency between fields that should be identical in different entries.

A bib file `mrabbrev.bib` containing string definitions for all the standard journal abbreviations used in *Mathematical Reviews* is part of $\mathcal{AMS}$-LaTeX and is also available from BibNet.

5. If you wish to include the URL of a file available on the Web, put it in a URL field. For example,

```
URL = "ftp://ftp.ma.man.ac.uk/pub/narep/narep306.ps.gz"
```

Although URL fields are not supported by the standard `bst` files, they are used in some BibNet databases and facilitate the creation of hypertext links from a `bib` file to the actual papers.

6. When writing a book or thesis you often wish to print out a draft chapter together with a bibliography comprising just the references cited in that chapter. This can be achieved using the LaTeX package `chapterbib` by Donald Arseneau, available from the CTAN archives.

Various tools are available to help maintain BIBTEX databases (see [109, §13.4] for details). These include tools to sort databases, search them, syntax check and pretty print them (the program `bibclean` [17] even checks ISBN and ISSN fields to see whether the checksum is correct), and extract from them the entries cited in a set of `aux` files (so as to create a `bib` file containing only the entries used in a particular document). Many of the utilities are available from BibNet. Some of these tasks are easy to do oneself using AWK, which is an interpreted programming language available on most Unix systems [4]. For example, the command line

```
awk 'BEGIN{ RS="" } /Riemann/' my.bib
```

searches the file `my.bib` in the current directory and prints all the entries that contain the word Riemann (assuming that records are separated by a blank line). Note that this AWK call prints complete `bib` entries, not just the lines on which the word occurs, as a `grep` search would.

13.4. Indexing and MakeIndex

An index to a book (or thesis, or report) has three main purposes.

1. To provide easy access to all the significant information.

2. To reveal relationships.

3. To reveal omissions.

A good index is therefore much more than a table of contents. But it is much less than a list of every important word, since it records useful information, not just key words. While a printed book is necessarily expressed in a linear order, the index is not constrained by ordering and can therefore reveal links between different parts of the book and bring together topics described in the text with varying terminology. A good index saves time for the reader as a result of what it *does not* contain: if a topic is not present in the index the reader can be sure that it is not covered in a significant way in the text.

An index should contain surprises—pointers to passages that the reader might overlook when scanning the book and its table of contents. It should anticipate the various ways in which a reader might search for a topic, by including it under multiple entries, where appropriate. For example, *block LU factorization* might be listed under *block, factorization, LU,* and, in a book not about matrices, under *matrix.* Since decomposition is a commonly used synonym for factorization, an entry "decomposition, *see* factorization" would also be appropriate in this example.

One source of index entries is section and subsection headings, since these provide the framework of the text. Entries should be nouns or nouns preceded by adjectives. Any conventions used in the index must be explained in a note at the beginning. For example, you might use "t" after a page number to denote reference to a table, and "f" for a figure.

If many names are to be indexed, it is worth creating separate name and subject indexes, as in this book. One reason for indexing names is to enable the reader to find where a particular paper in the bibliography is referenced, assuming, of course, that the author's name is mentioned at the point of reference.

A common mistake is to produce an index entry with too many page locators. If there are more than about five page locators, subentries should be introduced to help the reader pinpoint the information required. For example, the index entry

norm, 119–121, 123, 135, 159, 180

is much better broken down into, for example,

norm
        absolute, 119, 121
        dual, 120
        elliptic, 180
        Hölder inequality, 123
        spectral radius, relation with 135
        unitarily invariant, 159

In the following example the subentries serve little purpose because they all have the same page number:

LU factorization
        definition, 515
        existence, 515
        uniqueness, 515

This example should be collapsed into the single entry "LU factorization, 515", which is just as useful for a reader searching for information about the LU factorization.

Choose as main headings the word that the reader is most likely to look under. Thus

equations, displaying, 54

is better than

displaying equations, 54

This example is formatted as it should be if there are no other subentries of "equations". In the examples below, the subentry is assumed to be one of several and so appears on a separate line.

In subentries, use connectives to clarify the meaning of the entries. The entry

> slides
>> number, 138

could refer to a discussion on how to number the slides or on how many to produce. Adding the word "of" avoids the ambiguity:

> slides
>> number of, 138

It can be useful to add the word "of" even in unambiguous cases to make the entry read smoothly from subentry to heading:

> words
>> order of, 60

In traditional typesetting, indexing was a task to be done once a book was at the page-proof stage, and was often performed under severe time pressure. Nowadays, authors typesetting their own books by computer can index earlier in the production process, making use of indexing software.

MakeIndex is a C program, written by Pehong Chen [56], [109, Chap. 12] with advice from Leslie Lamport, that makes an index for a LATEX document. The user has to place `\index` commands in the LATEX source that define the name and location of the items to be indexed. If a `\makeindex` command is placed in the preamble (before `\begin{document}`) then LATEX writes the index entries, together with the page numbers on which they occur, to an `idx` file. This is read by the MakeIndex program, which processes and sorts the information, producing an `ind` file that generates the index when included in the LATEX document (see Figure 13.1). MakeIndex provides various options in the `\index` command to support standard indexing requirements, such as subentries, page ranges and cross-references to other entries. Here is how the beginning of one sentence from page 187 was typed:

```
\item  It is easy to prepare
transparencies\index{slides!preparing in \TeX}
with \TeX\ if the paper
```

The exclamation mark in the `\index` command denotes the beginning of a subentry. Multiple indexes (such as name and subject indexes) can be produced with the aid of the `index` package by David M. Jones, available from `ftp://theory.lcs.mit.edu/pub/tex/index/`

Here are some tips on indexing in LATEX.

1. Insert the index entry immediately following the word to be indexed, on the same line and with no spaces before the \index command (as in the example above). This ensures that the correct page reference is produced and avoids unwanted spaces appearing in the output.

2. If the scope of the item being indexed is more than one sentence, so that the scope may be broken over a page, index the item as a page range. For example, this list of tips is contained within commands

   ```
   \index{LaTeX@\LaTeX!indexing in|(}
   \index{indexing!in latex@in \LaTeX|(}
   ...
   \index{indexing!in latex@in \LaTeX|)}
   \index{LaTeX@\LaTeX!indexing in|)}
   ```

   The |(and |) strings serve to delimit the range of the index command.

3. *See* entries can be produced by commands of the form

   ```
   \index{dots|see{ellipsis}}
   ```

   To produce *see also* entries in an analogous way you can use the following definition, adapted from that for \see in makeidx.sty:

   ```
   \newcommand\seealso[2]{\emph{see also} #1}
   ```

   Place all *see* and *see also* index entries together, to make it easier to edit them and check for consistency. I suggest placing them after the last item in the book to be indexed (ideally, just before the bibliography); this ensures that *see also* appears after the page references for an entry.

4. Do not leave the task of indexing to the very last stage. For, in inserting the \index entries, you are likely to introduce errors (of spacing, at least) and so a further round of proofreading will be needed after the indexing stage.

AWK tools for indexing are described in [4, §5.3] and [22]; these tools do not support subentries. A simple and elegant way to construct key word in context (KWIC) indexes using AWK is also described in [4]. A KWIC index lists each word in the context of the line in which it is found; the list is sorted by word and arranged so that the key words line up. One of the main uses of KWIC indexes is to index titles of papers.

For an interesting example of an index, see Halmos's *I Want to Be a Mathematician* [127]. Priestley [230] says "If index writing has not bloomed into an art form in *I Want to Be a Mathematician*, it has at least taken a quantum leap forward. From 'academic titles, call me mister' to 'Zygmund, A., at faculty meetings' this one is actually worth reading."

13.5. Further Sources of Information

The best (and the most humorous) introduction to LATEX is *Learning LATEX* [118] by Griffiths and D. J. Higham. A much longer and more detailed book that is very handy for reference is *A Guide to LATEX 2_ε* [166] by Kopka and Daly. Lamport's *LATEX: A Document Preparation System* [172] is the "official" guide to LATEX. For those still using the obsolete LATEX 2.09, Carlisle and Higham [53] explain the advantages to be gained by upgrading to LATEX 2_ε.

For technical details of LATEX, BIBTEX and MakeIndex, and descriptions of the many available packages, see *The LATEX Companion* [109] by Goossens, Mittelbach and Samarin. *The LATEX Graphics Companion* [110] by Goossens, Rahtz and Mittelbach is the most comprehensive and up-to-date reference on producing graphics with LATEX and PostScript. If you are a really serious LATEX or TEX user you will want to study Knuth's *The TEXbook* [161], the "bible" of TEX, or another advanced reference such as Salomon's *The Advanced TEXbook* [244].

BIBTEX is described in all the LATEX textbooks mentioned above, but most comprehensively in *The LATEX Companion* [109].

An article by Knuth [158][24] offers many insights into mathematical typesetting and type design, and describes early versions of TEX and METAFONT (METAFONT [160] is Knuth's system for designing typefaces).

The Comprehensive TEX Archive Network (CTAN) is a network of ftp servers that hold up-to-date copies of all the public domain versions of TEX, LATEX, and related macros and programs. The three main sites are at

```
ftp.dante.de,    http://www.dante.de/
ftp.tex.ac.uk,   http://www.tex.ac.uk/tex-archive
tug2.cs.umb.edu, http://tug2.cs.umb.edu/ctan/
```

which are located in Germany, England and Massachusetts, USA, respectively. There are many mirror sites around the world, details of which may be obtained from the TEX Users Group (TUG) Web pages. The organization of TEX files is the same on each site and starts at `./tex-archive`. To search a CTAN site during an anonymous ftp session type the command

---

[24]The beginning of the abstract is quoted on page 86.

`quote site index string`, where `string` is a Unix regular expression (a filename optionally containing wildcards) on which to search.

The TUG runs courses and conferences on TeX and produces a journal called *TUGboat*. It also produces a newsletter for members called *TeX and TUG News*. Contact details for TUG are given in Appendix D.

The UK TeX Users Group, based in the UK, also organizes meetings and produces a newsletter (called *Baskerville*). It cooperates with TUG and supports the UK TeX archive (the UK node of CTAN). More information is available on the Web at `http://www.tex.ac.uk/UKTUG/` or via email to `uktug-enquiries@uk.ac.tex`

Excellent advice on preparing an index is given in *The Chicago Manual of Style* [58] and in Bonura's *The Art of Indexing* [35]. The collection *Indexers on Indexing* [130] contains articles on many different aspects of indexing that originally appeared in *The Indexer*, the journal of the Society of Indexers (UK). This society is involved in awarding indexing prizes; the 1975 Wheatley Medal was awarded for the index (by Margaret D. Anderson) to the first edition of [45]. Other good references are *Words into Type* [249] and *Copy-Editing* [45].

Chapter 14
Aids and Resources for Writing and Research

> The library is the mathematician's laboratory.
> — PAUL R. HALMOS, *I Want to be a Mathematician:*
> *An Automathography in Three Parts* (1985)

> *It's a library, honey—kind of an early version of the World Wide Web.*
> — From a cartoon by ED STEIN

> *Once you master spelling* anonymous,
> *you can roam around the public storage areas*
> *on computers on the Internet*
> *just as you explore public libraries.*
> — TRACY LAQUEY and JEANNE C. RYER, *The Internet Companion* (1993)

> *Just as footnotes and a bibliography trace an idea's ancestors,*
> *citation indexing traces an idea's offspring.*
> — KEVIN KELLY, in *SIGNAL: Communication*
> *Tools for the Information Age* (1988)

14.1. Internet Resources

A huge variety of information and software is available over the Internet, the worldwide combination of interconnected computer networks. The location of a particular object is specified by a URL, which stands for "Uniform Resource Locator". Examples of URLs are

```
http://www.netlib.org/index.html
ftp://ftp.netlib.org
```

The first example specifies a World Wide Web server (http = hypertext transfer protocol) together with a file in hypertext format (html = hypertext markup language), while the second specifies an anonymous ftp (file transfer protocol) site. In any URL, the site address may, optionally, be followed by a filename that specifies a particular file.

The best way of accessing information on the Internet is with a World Wide Web browser, such as Netscape Navigator or Microsoft Internet Explorer. These browsers have intuitive interfaces, making them very easy to learn. For downloading files an alternative is to use an ftp program to carry out anonymous ftp. Anonymous ftp is a special form of ftp in which you log on as user **anonymous** and need not type a password (though, by convention, you are supposed to type your email address to indicate who you are). Table 14.1 lists some of the file types you may encounter when ftp'ing files.

For more details about the Internet and how to access it see on-line information, or one of the many books on the subject, such as Krol [168].

Newsgroups

The news system available on many computer networks contains a large number of newsgroups to which users contribute messages. The newsgroups frequently carry announcements of new software and software updates. On a Unix system, type **man rn** or **man nn** for information on how to read news. Newsgroups of general interest to mathematicians are **sci.math** and its more specialized cousins such as **sci.math.research** and **sci.math.symbolic**, and **comp.text.tex** for TeX information. For many newsgroups a FAQ document of Frequently Asked Questions is available.

Digests

Various magazines are available by email. These collect questions, answers and announcements submitted to a particular email address. For example, NA-Digest is a weekly magazine about numerical analysis [73]; send mail to **na.help@na-net.ornl.gov** for information on how to subscribe.

Table 14.1. Standard file types.

Suffix	Type	Explanation
.bib	ASCII	BibTeX source.
.bst	ASCII	BibTeX style file.
.dvi	binary	TeX output. Use dvips to convert to Post-Script.
.gif, .tif	binary	Image file formats (Graphics Interchange Format, Tagged Image File Format) [37].
.gz	binary	Compressed. Use gunzip or gzip -d to recover original file.
.ist	ASCII	MakeIndex style file.
.pdf	ASCII	Portable Document Format (PDF), developed by Adobe Systems, Inc. Can be read using the Adobe Acrobat software.
.ps	ASCII	PostScript file.
.shar	ASCII	"Shell archive" collection of files. Use sh to extract files.
.sty	ASCII	LaTeX style file.
.tar	binary	"Tape archive" collection of files. Use tar -xvf or pax -r to extract the contents.
.tex	ASCII	TeX source.
.txt	ASCII	Text file.
.uu	ASCII	Re-coded form of binary file, suitable for mailing. Use uudecode to recover original binary file.
.Z	binary	Compressed. Use uncompress to recover original file.
.z	binary	Compressed by an older algorithm. Use unpack to recover original file.
.zip	binary	Compressed by PKZIP. Use an unzip program to recover original file.

Netlib

Netlib is an electronic repository of public domain mathematical software for the scientific computing community. In particular, it houses the various -PACK program libraries, such as EISPACK, LINPACK, MINPACK and LAPACK, and the collected algorithms of the Association for Computing Machinery (ACM). Netlib has been in existence since 1985 and can be accessed by email, ftp or the Web. In addition to providing mathematical software, netlib provides the facility to download technical reports from certain institutions, to download software and errata for textbooks, and to search the SIAM membership list (via the `whois` command). Background on netlib is given in an article by Dongarra and Grosse [72] (see also [40]) and news of the system is published regularly in NA-Digest and *SIAM News* (received by every personal and institutional member of SIAM). To obtain a catalogue of the contents of netlib send an email message with body `send index` to `netlib@ornl.gov` or `netlib@research.att.com`. Alternatively, netlib can be accessed over the Web at the address `http://www.netlib.org/index.html`. Copies of netlib exist at various other sites throughout the world.

e-MATH

The AMS runs a computer service, e-MATH, with many features. For example, it allows you to obtain pointers to reviews in *Mathematical Reviews* (1985 to present), to download the list of Mathematics Subject Classifications, to download articles (in one of several formats, including `dvi`, PostScript and TEX) from the *Bulletin of the American Mathematical Society* and other journals, to access lists of employment and post-doctoral opportunities, and to search the combined membership list of the AMS, the Mathematical Association of America (MAA), SIAM, and the American Mathematical Association of Two-Year Colleges (AMATYC). E-MATH is best accessed via its Web interface at `http://www.ams.org/`. Status reports on e-MATH are published in the *Notices of the American Mathematical Society* in the "Inside the AMS" column (every personal and institutional member of the AMS receives this journal).

14.2. Library Classification Schemes

The two main classification schemes used in libraries are the Library of Congress Classification and the Dewey Decimal Classification.

The Library of Congress Classification was developed in the early 1900s for the collections of the Library of Congress in the US. The main classes are denoted by single capital letters, the subclasses by two capital letters, and divisions of the subclasses by integers, which themselves can be subdivided beyond the decimal point. Mathematics is subclass QA of the science class Q; an outline of this subclass is given in Table 14.2. Every book is identified by a call number. For example, the first edition of this book has the call number QA42.H54, where 42 is the subdivision of class QA described as "Communication of mathematical information, language, authorship" and H54 is the author number.

The Dewey Decimal Classification was first introduced in 1876 and is used by most libraries in the UK. It divides knowledge into ten different broad subject areas called classes, numbered 000, 100, ..., 900. Class 500 covers Natural Sciences and Mathematics, and subclass 510 Mathematics. Table 14.3 gives an outline of subclass 510.

Both schemes were developed to classify the mathematics of the nineteenth century, so some modern areas of mathematics fit into them rather awkwardly. Variation is possible in the way the schemes are used in different libraries. For example, in the John Rylands University Library of Manchester the unassigned sections 517 and 518 are used for analysis and numerical analysis, respectively. My experience is that because of the vagaries of the schemes and the differing opinions of librarians who choose classifications, books are often not classified in the way you would expect. If you are looking for a specific book, searching for it by author and title (or by ISBN) in an on-line catalogue is usually the best way of locating it.

14.3. Review, Abstract and Citation Services

When you need to find out what work has been done in a particular area or by a particular author, or need to track down an incomplete reference, you should consult one of the reviews or citation collections. The main ones are as follows.

Mathematical Reviews (MR) is run by the American Mathematical Society (AMS) and was first published in 1940. Each month the journal publishes short reviews of recently published papers drawn from approximately 2000 scholarly publications. Each review either is written by one of the nearly 12,000 reviewers or is a reprint of the paper's abstract. The reviews are arranged in accordance with the Mathematics Subject Classifications (see §6.7). MR is particularly useful for finding details of a paper in a journal that your library does not receive—based on the review you can decide whether to order the paper via the inter-library service. Sometimes you will see an entry in a reference list containing a term such as MR 31 #1635. This means that the article in question was reviewed in volume 31

Table 14.2. Outline of Library of Congress Classification subclass QA.

1–99	General mathematics
101–145	Elementary mathematics, arithmetic
150–272	Algebra
273–274	Probabilities
276–280	Mathematical statistics
281	Interpolation
292	Sequences
295	Series
297–299	Numerical analysis
300–433	Analysis
401–433	Analytical problems used in the solution of physical problems
440–699	Geometry (including topology)
801–939	Analytical mechanics

Table 14.3. Outline of Dewey Decimal Classification subclass 510 (21st edition, 1996).

510^a	Mathematics
511	General principles of mathematics
512	Algebra, number theory
513	Arithmetic
514	Topology
515	Analysis
516	Geometry
517	Unassigned
518	Unassigned
519	Probabilities and applied mathematics

[a]Section 510 covers the general works of the entire subclass.

of MR as review number 1635.

The AMS also produces *Current Mathematical Publications* (CMP), which is essentially a version of MR that contains the bibliographic records but not the reviews. However, CMP is much more up to date than MR: it is issued every three weeks and contains a list of items received by the MR office, most of which will eventually be reviewed in MR.

Computing Reviews (CR) plays a role for computer science similar to the one MR plays for mathematics. It is published by the Association for Computing Machinery and uses its own classification scheme (see §6.7). The other major abstracting journal for computer science is *Computer and Control Abstracts*; it has wider coverage than CR.

Current Contents (CC), from the Institute for Scientific Information (ISI), Philadelphia, is a weekly list of journal contents pages (similar in size to the US *TV Guide*). The Physical Sciences edition is the one in which mathematics and computer science journals appear. Each issue of CC is arranged by subject area and contains an index of title words. Each issue also contains an article by Eugene Garfield, the founder of ISI; these articles often report citation statistics, such as most-cited papers in particular subject areas.

The *Science Citation Index* (SCI), also from the ISI, records reference lists of papers in such a way that the question "which papers cite a given paper?" can be answered. Approximately 3300 journals are indexed at present, across all science subjects (the total number of scholarly science journals is of the order 25,000). The SCI began in the early 1960s and covers the period from 1945 to the present [88], [91]. The SCI provides a means for finding newer papers that were influenced by older ones, whereas searching reference lists takes you in the opposite direction. For example, suppose we are interested in the reference

W. KAHAN, *Further remarks on reducing truncation errors*, Comm. ACM, 8 (1965), p. 40.

If we look under "KAHAN W" in the five-year cumulation 1975–1979 *Citation Index*, and then under the entry for his 1965 paper, we find four papers that include Kahan's in their reference lists. For each of these citing papers the first author, journal, volume, starting page number and year of publication are given. The full bibliographic data for these papers can be found in the SCI *Source Index*. Looking up these four papers in later indexes we find further references on the topic of interest.

If you can remember the title of a paper but not the author, the SCI *Permuterm Subject Index* can help. This is a key word index in which every significant word in each article title is paired with every other significant word from the same title. Under each pair of key words is a list of relevant authors; their papers may be found in the *Source Index*. As an example, if all we know of Kahan's article are the words "truncation" and "errors" and the year of publication we can find the full details of the article from the five-year cumulation 1965–1969 *Permuterm Subject Index* and the corresponding *Source Index*. Garfield's article "How to Use the Science Citation Index" [99] gives detailed examples of the use of the SCI and is well worth reading.

Electronic versions of the SCI can be used to search for papers with given key words in the title, abstract or indexing fields, to obtain a list of papers by an author, and to obtain a list of an author's cited works, showing the number of times and where each work has been cited. A limitation of the SCI is that it records only the first author of a cited paper, so citations to a paper by "Smith and Jones" will benefit Smith's citation count but not Jones's.

The ISI has a Web page at `http://www.isinet.com/`. It contains some of Garfield's past articles about citation indexing and gives details about electronic access to the ISI products.

Zentralblatt für Mathematik und ihre Grenzgebiete (ZM), also titled *Mathematical Abstracts*, is another mathematical review journal. It was founded in 1931 and is published by Springer-Verlag and Fachinformationszentrum Karlsruhe. It uses the Mathematics Subject Classifications, and its coverage is almost identical to that of MR.

MR, SCI and ZM are available in computer-readable form on compact disc (CD-ROM) and in on-line databases.

A number of databases, including the ISI Citation Indexes, can be accessed from BIDS (Bath Information and Data Services) at `http://www.bids.ac.uk`, which operates from the University of Bath in the UK. Most of the databases are available on an institutional-license basis, with most UK universities having a license.

14.4. Text Editors

As the use of computers in research and writing increases, we spend more and more time at the keyboard, much of it spent typing text with a text editor. Of all the various programs we use, the text editor is the one that generates the most extreme feelings: most people have a favourite editor and fervently defend it against criticism. Under the Unix operating system two editors are by far the most widely used. The first, and oldest, is `vi`, which has the advantage that it is available on every Unix system. The second, and the more powerful, is Emacs. Not only does Emacs do almost everything you would expect of a text editor, but from within it you can run other programs and view and edit their output; rename, move

---

Citation Facts and Figures

The *Science Citation Index* (1945–1988) contains about 33 million cited items. The most-cited paper is

> O. H. Lowry, N. J. Rosebrough, A. L. Farr and R. J. Randall, Protein measurement with the Folin phenol reagent, *J. Biol. Chem.*, 193:265–275, 1951.

which has 187,652 citations. The next most cited paper (also on protein methods, as are all of the top three) has 59,759 citations. The six most-cited papers from Mathematics, Statistics and Computer Science are as follows; their rankings on the list of most-cited papers range from 24th to 297th.

1. D. B. Duncan, Multiple range and multiple F tests, *Biometrics*, 11:1–42, 1955. (8985 citations)

2. E. L. Kaplan and P. Meier, Nonparametric estimation from incomplete observations, *J. Amer. Statist. Assoc.*, 53:457–481, 1958. (4756 citations)

3. D. W. Marquardt, An algorithm for least-squares estimation of nonlinear parameters, *J. Soc. Indust. Appl. Math.*, 11:431–441, 1963. (3441 citations)

4. D. R. Cox, Regression models and life-tables, *J. Royal Statist. Soc. Ser. B Metho.*, 34:187–220, 1972. (3392 citations)

5. R. Fletcher and M. J. D. Powell, A rapidly convergent descent method for minimization, *Comp. J.*, 6:163–168, 1963. (1948 citations)

6. J. W. Cooley and J. W. Tukey, An algorithm for the machine calculation of complex Fourier series, *Math. Comp.*, 19:297–301, 1965. (1845 citations)

In the period 1945–1988, 55.8% of the papers in the index were cited only once, and 24.1% were cited 2–4 times. 5767 papers were cited 500 or more times. References: [93], [94], [96], [97], [98].

and delete files; send and read electronic mail; and surf the Web. Some workstation users carry out nearly all their computing "within Emacs", leaving it running all the time.

There are various versions of Emacs, one of the most popular of which is GNU Emacs [51], [107]. (GNU stands for "Gnu's not UNIX" and refers to a Unix-like operating system that is being built by Richard Stallman and his associates at the Free Software Foundation.) GNU Emacs is available for workstations and 386 (and above)-based PC-compatibles; other PC versions include Freemacs, MicroEmacs and Epsilon. GNU Emacs contains modes for editing special types of files, such as TEX, LATEX and Fortran files. In these modes special commands are available; for example, in TEX mode one Emacs command will invoke TEX on the file in the current editing buffer. Appendix C contains a list of the 60+ most useful GNU Emacs commands and should be helpful to beginners and intermediate users.

For anyone who spends a lot of time typing at a computer, learning to touch type is essential. This need not be time-consuming, and can be done using one of the self-tutor programs available.

14.5. Spelling Checking, Filters and Pipes

Programs that check, and possibly correct, spelling errors are widely available. These are useful tools, even for the writer who spells well, because, as McIlroy has observed [199], most spelling errors are caused by errors in typing. The Unix[25] spelling checker `spell` takes as input a text file and produces as output a list of possibly misspelled words. The list comprises words that are not in `spell`'s dictionary and which cannot be generated from its entries by adding certain inflections, prefixes or suffixes; a special "stop list" avoids non-words such as *beginer* (*begin* + *er*) being accepted. The development of `spell` is described in a fascinating article by McIlroy [199] and is summarized by Bentley in [20, Chap. 13]. When I ran an earlier version of this book through `delatex` (see below), followed by `spell` with the British spelling option, part of the output I obtained was as follows (the output is wrapped into columns here to save space)

bbl	capitalized	de	diag
beginer	cccc	deci	dispensible
blah	co	delatex	dvi
blocksize	comp	dependant	ees
bst	computerized	der	eg

---

[25]More details on the Unix commands described in this section can be obtained from a Unix reference manual or from the on-line manual pages by typing `man` followed by the command name.

The -ize endings are flagged as errors because `spell` expects you to use -ise endings when the British spelling option is in effect. The one genuine error revealed by this extract is *dispensible*, which should be *dispensable*. `Spell` can be instructed to remove from its output words found in a supplemental list provided by the user. This way you can force `spell` to accept technical terms, acronyms and so on. `Spell` can be used to check a single word, *mathematical*, say, by typing at the command line

```
spell
mathematical
^d
```

(`^d` is obtained by holding down ctrl and typing d). In this case there is no output because the word is recognized by `spell`.

If you use GNU Emacs, you can call the Unix `spell` program from within the editor. The command `Esc-$` checks the word under the cursor and `Esc-x spell-buffer` checks the spelling of the whole buffer. In each case you are given the opportunity to edit an unrecognized word and replace all occurrences with the corrected version.

A problem with spell checking TEX documents is that most TEX commands will be flagged as errors. When working in Unix the solution is to run the file through `detex`, or `delatex`[26] for LaTeX documents, before passing it to `spell`; these filters strip the file of all TEX and LaTeX commands, respectively. An alternative is simply to have the spelling checker learn the TEX and LaTeX command names as though they were valid words.

It is important to realize that spelling checkers will not identify misspellings that are themselves words, such as *form* for *from* (a common error in published papers), *except* for *expect*, or *conversation* for *conversation*. For this reason, bigger does not necessarily mean better for dictionaries used by spelling checkers. Peterson [225] investigates the relationship between dictionary size and the probability of undetected typing errors; he recommends that "word lists used in spelling programs should be kept small."

Spelling correction programs are also available on various computers. These not only flag unrecognized words, but present guesses of the correct spelling. They look for errors such as transposition of letters, or a single letter extra, missing or incorrect (research is mentioned by Peterson in [224] that finds these to be the cause of 80% of all spelling errors). The suggested corrections can be amusing: for example, one spelling corrector I have used suggests dunce for Duncan and turkey for Tukey.

Ispell is an interactive spelling checker and corrector available for Unix and DOS systems. When invoked with a filename it displays each word

---

[26]This filter is available from netlib in the `typesetting` directory (see §14.1).

in the file that does not appear in its dictionary and offers a list of "near misses" and guesses of the correct word. You can accept one of the suggested words or type your own replacement. An Ispell interface exists for GNU Emacs.

You can do limited searching of spell's dictionary using the look command, which finds all words with a specified prefix. Thus

```
look comp | grep ion$
```

displays all words that begin with *comp-* and end in *-ion*.

It is interesting to examine the frequency of word usage in your writing. Under Unix this can be done with the following pipe, where file is a filename:

```
cat file | deroff -w | tr A-Z a-z | sort | uniq -c | sort -rn
```

The deroff -w filter divides the text into words, one per line (at the same time removing any troff commands that may be present), tr A-Z a-z converts all words to lower case, sort sorts the list, uniq -c converts repeated lines into a single line preceded by a count of how many times the line occurred, and sort -rn sorts on the numeric count field in reverse order (largest to smallest). Applying this pipe to an earlier version of this book I obtained as the first twenty-five lines of output (wrapped into columns here to save space)

1533 the	529 and	266 for	185 by	154 i
709 a	517 in	259 that	175 as	154 you
696 of	310 ndex	230 are	167 egin	150 not
690 to	275 it	220 mph	164 nd	146 or
613 is	267 be	186 ite	160 this	141 with

(The non-words ndex, mph, ite, egin, and nd are left-over LATEX commands, which appear since I did not use delatex in the pipe.) It is worth examining word frequency counts to see if you are overusing any words. As far as I am aware, this particular count does not reveal any abnormalities in my word usage.

The Unix diff command takes two files and produces a list of changes that would have to be made to the first file to make it the same as the second. The changes are expressed in a syntax similar to that used in the ed text editor. If you use the -c option (diff -c filename) then the three lines before and after each change are printed to show the context. The main use of diff for the writer is to see how two versions of a file (a current and an earlier draft, say) differ. If your co-author updates the source file for a TEX paper, you can use diff to see what changes have been made.

Another useful filter is the `wc` command, which counts the lines, words and characters in a file. When I ran the source for an almost-final draft of this book through `wc` (using the command `wc *.tex`, since the source is contained in several `.tex` files) I obtained as the final line of output

```
17312    87478   647544    total
```

This count shows that the source contains 87,478 words, though many of these words are LaTeX instructions that do not result in a printed word, so this is an overestimate of the actual word count. When I ran the source through `delatex` before sending it to `wc` the word count dropped to 73,302.

14.6. Style Checkers

Programs exist that try to check the style of your text. Various commercial programs are available for PCs. Style checking programs have been available for Unix machines since the late 1970s. An article by Cherry [57] describes several programs: these include `style`, which "reads a document and prints a summary of readability indices, sentence length and type, word usage, and sentence openers", and `diction`, which "prints all sentences in a document containing phrases that are either frequently misused or indicate wordiness".

One of the readability formulas used by `style` is the Kincaid formula, which assigns the reading grade (relative to the US schooling system)

$$11.8 \times \text{(average syllables per word)} + 0.39 \times \text{(average words per sentence)}$$
$$- 15.59.$$

The formula was derived by a process that involved measuring how well a large sample of US Navy personnel understood Navy technical manuals. (It is a contractual requirement of the US Department of Defense that technical manuals achieve a particular reading measure.)

In his book *The Art of Plain Talk* [81, Chap. 7], Flesch proposes the following score for measuring the difficulty of a piece of writing:

$$s = 0.1338 \times \text{(average words per sentence)}$$
$$+ 0.0645 \times \text{(average affixes per 100 words)}$$
$$- 0.0659 \times \text{(average personal references per 100 words)} - 0.75.$$

The score is usually between 0 and 7, and Flesch classifies the scores in unit intervals from $s \leq 1$, "very easy", to $s > 6$, "very difficult". He states that comics fall into the "very easy" class and scientific journals into the "very difficult" class. Klare [155] explains that "Prior to Flesch's time, readability was a little-used word except in educational circles, but he made it an important concept in most areas of mass communication."

The limitations of readability indices are well known and are recognized by their inventors [155], [198]. For example, the Kincaid and Flesch formulas are invariant under permutations of the words of a sentence. More generally, readability formulas measure style and not clarity, content or organization. For the writer, a readability formula is best regarded as a rough means for rating a draft to see whether it is suitable for the intended audience.

AT&T's Bell Laboratories markets *The Writer's Workbench*, an extensive system that incorporates `style`, `diction` and various other programs [186]. One of these is `double`, which checks for occurrences of a word twice in succession, possibly on different lines. Repeated words are hard for a human proofreader to detect.

Hartley [133] obtained suggestions from nine colleagues about how to improve a draft of his paper [132] and compared them with the suggestions generated by *The Writer's Workbench*. He concluded that "Text-editing programs can deal well with textual issues (perhaps better than humans) but humans have prior knowledge and expertise about content which programs currently lack."

Knuth ran a variety of sample texts through `diction` and `style` [164, §40], including technical writing, Wuthering Heights and Grimm's Fairy Tales. He found that his book of commentaries on Chapter 3, verse 16 of each book of the Bible [162] was given a significantly lower reading grade level than the other samples, and concluded that we tend to write more simply when writing outside our own field.

Appendix A
The Greek Alphabet

Capital	Lower case	English name
A	α	alpha
B	β	beta
Γ	γ	gamma
Δ	δ	delta
E	ϵ, ε	epsilon
Z	ζ	zeta
H	η	eta
Θ	θ, ϑ	theta
I	ι	iota
K	κ	kappa
Λ	λ	lambda
M	μ	mu
N	ν	nu
Ξ	ξ	xi
O	o	omicron
Π	π	pi
P	ρ, ϱ	rho
Σ	σ, ς	sigma
T	τ	tau
Υ	υ	upsilon
Φ	ϕ, φ	phi
X	χ	chi
Ψ	ψ	psi
Ω	ω	omega

Appendix B
Summary of TEX and LATEX Symbols

This appendix is based on `symbols.tex` version 3.2 by David Carlisle, available from the CTAN Archives.

Table B.1. Accents.

ò	\`{o}	õ	\~{o}	ǒ	\v{o}	ǫ	\c{o}	ó	\'{o}
ō	\={o}	ő	\H{o}	ọ	\d{o}	ô	\^{o}	ȯ	\.{o}
o͡o	\t{oo}	o̲	\b{o}	ö	\"{o}	ŏ	\u{o}	o̊	\r{o}

Table B.2. Dotless letters for use with accents.

ı	\i	ȷ	\j

Table B.3. Math mode accents.

ä	\ddot{a}	á	\acute{a}	ā	\bar{a}	ȧ	\dot{a}
ă	\breve{a}	ǎ	\check{a}	à	\grave{a}	ā⃗	\vec{a}
â	\hat{a}	â	\widehat{a}	ã	\tilde{a}	ã	\widetilde{a}

Table B.4. Foreign symbols.

œ	\oe	Œ	\OE	æ	\AE	Æ	\AE	å	\aa
Å	\AA	ø	\o	Ø	\O	ł	\l	Ł	\E
ß	\ss	SS	\SS	¿	?`	¡	!`		

225

Table B.5. Greek letters.

α	\alpha	θ	\theta	o	o	τ	\tau
β	\beta	ϑ	\vartheta	π	\pi	υ	\upsilon
γ	\gamma	γ	\gamma	ϖ	\varpi	ϕ	\phi
δ	\delta	κ	\kappa	ρ	\rho	φ	\varphi
ϵ	\epsilon	λ	\lambda	ϱ	\varrho	χ	\chi
ε	\varepsilon	μ	\mu	σ	\sigma	ψ	\psi
ζ	\zeta	ν	\nu	ς	\varsigma	ω	\omega
η	\eta	ξ	\xi				

Γ	\Gamma	Λ	\Lambda	Σ	\Sigma	Ψ	\Psi
Δ	\Delta	Ξ	\Xi	Υ	\Upsilon	Ω	\Omega
Θ	\Theta	Π	\Pi	Φ	\Phi		

Table B.6. Binary operation symbols.

$\pm$	\pm	$\cap$	\cap	$\diamond$	\diamond
$\mp$	\mp	$\cup$	\cup	$\triangle$	\bigtriangleup
$\times$	\times	$\uplus$	\uplus	$\triangledown$	\bigtriangledown
$\div$	\div	$\sqcap$	\sqcap	$\triangleleft$	\triangleleft
$*$	\ast	$\sqcup$	\sqcup	$\triangleright$	\triangleright
$\star$	\star	$\vee$	\vee (or \lor)	$\lhd$	\lhd*
$\circ$	\circ	$\wedge$	\wedge (or \land)	$\rhd$	\rhd*
$\bullet$	\bullet	$\setminus$	\setminus	$\unlhd$	\unlhd*
$\cdot$	\cdot	$\wr$	\wr	$\unrhd$	\unrhd*
$\oplus$	\oplus	$\ominus$	\ominus	$\otimes$	\otimes
$\oslash$	\oslash	$\odot$	\odot	$\bigcirc$	\bigcirc
$\dagger$	\dagger	$\ddagger$	\ddagger	$\amalg$	\amalg
$+$	+	$-$	-		

* Not predefined in LATEX 2_ε. Use one of the packages latexsym, amsfonts or amssymb.

Table B.7. Punctuation symbols.

,	,	;	;	:	\colon	.	\ldotp	·	\cdotp

Table B.8. Relation symbols.

$\leq$	`\leq` (or `\le`)	$\geq$	`\geq` (or `\ge`)	$\equiv$	`\equiv`
$\prec$	`\prec`	$\succ$	`\succ`	$\sim$	`\sim`
$\preceq$	`\preceq`	$\succeq$	`\succeq`	$\simeq$	`\simeq`
$\ll$	`\ll`	$\gg$	`\gg`	$\asymp$	`\asymp`
$\subset$	`\subset`	$\supset$	`\supset`	$\approx$	`\approx`
$\subseteq$	`\subseteq`	$\supseteq$	`\supseteq`	$\cong$	`\cong`
$\sqsubset$	`\sqsubset`*	$\sqsupset$	`\sqsupset`*	$\neq$	`\neq` (or `\ne`)
$\sqsubseteq$	`\sqsubseteq`	$\sqsupseteq$	`\sqsupseteq`	$\doteq$	`\doteq`
$\in$	`\in`	$\ni$	`\ni` (or `\owns`)	$\notin$	`\notin`
$\vdash$	`\vdash`	$\dashv$	`\dashv`	$<$	`<`
$\models$	`\models`	$\perp$	`\perp`	$\mid$	`\mid`
$\parallel$	`\parallel`	$\bowtie$	`\bowtie`	$\bowtie$	`\Join`*
$\smile$	`\smile`	$\frown$	`\frown`	$\propto$	`\propto`
$=$	`=`	$>$	`>`	$:$	`:`

* Not predefined in LATEX 2_ε. Use one of the packages `latexsym`, `amsfonts` or `amssymb`.

These symbols can be negated by preceding them with `\not`.
Example: `\not\equiv` produces $\not\equiv$.

Table B.9. Arrow symbols.

$\leftarrow$	`\leftarrow` (or `\gets`)	$\longleftarrow$	`\longleftarrow`
$\Leftarrow$	`\Leftarrow`	$\Longleftarrow$	`\Longleftarrow`
$\rightarrow$	`\rightarrow` (or `\to`)	$\longrightarrow$	`\longrightarrow`
$\Rightarrow$	`\Rightarrow`	$\Longrightarrow$	`\Longrightarrow`
$\leftrightarrow$	`\leftrightarrow`	$\longleftrightarrow$	`\longleftrightarrow`
$\Leftrightarrow$	`\Leftrightarrow`	$\Longleftrightarrow$	`\Longleftrightarrow`[†]
$\mapsto$	`\mapsto`	$\longmapsto$	`\longmapsto`
$\hookleftarrow$	`\hookleftarrow`	$\hookrightarrow$	`\hookrightarrow`
$\leftharpoonup$	`\leftharpoonup`	$\rightharpoonup$	`\rightharpoonup`
$\leftharpoondown$	`\leftharpoondown`	$\rightharpoondown$	`\rightharpoondown`
$\rightleftharpoons$	`\rightleftharpoons`	$\leadsto$	`\leadsto`*
$\uparrow$	`\uparrow`	$\downarrow$	`\downarrow`
$\Uparrow$	`\Uparrow`	$\Downarrow$	`\Downarrow`
$\updownarrow$	`\updownarrow`	$\Updownarrow$	`\Updownarrow`
$\nwarrow$	`\nwarrow`	$\nearrow$	`\nearrow`
$\swarrow$	`\swarrow`	$\searrow$	`\searrow`

* Not predefined in LATEX 2_ε. Use one of the packages `latexsym`, `amsfonts` or `amssymb`.
[†] Also `\iff`, which puts an extra thick space on either side.

Table B.10. Miscellaneous symbols.

†	\dag	§	\S	©	\copyright
‡	\ddag	¶	\P	£	\pounds*
LaTeX	\LaTeX*	LaTeX 2ε	\LaTeXe*		
...	\ldots#	⋯	\cdots	⋮	\vdots
⋰	\ddots				
ℵ	\aleph	′	\prime	∀	\forall
ℏ	\hbar	∅	\emptyset	∃	\exists
ı	\imath	∇	\nabla	¬	\neg (or \lnot)
ȷ	\jmath	√	\surd	♭	\flat
ℓ	\ell	⊤	\top	♮	\natural
℘	\wp	⊥	\bot	♯	\sharp
ℜ	\Re	\	\backslash	∠	\angle
ℑ	\Im	∂	\partial	℧	\mho*
.	.	∞	\infty	□	\Box*
◇	\Diamond*	△	\triangle	♣	\clubsuit
◇	\diamondsuit	♡	\heartsuit	♠	\spadesuit

* Predefined in LATEX 2ε but not in TEX.
# Or \dots.

Table B.11. Variable-sized symbols.

∑	\sum	∩	\bigcap	⊙	\bigodot
∏	\prod	∪	\bigcup	⊗	\bigotimes
∐	\coprod	⊔	\bigsqcup	⊕	\bigoplus
∫	\int	⋁	\bigvee	⊎	\biguplus
∮	\oint	⋀	\bigwedge		

Table B.12. Log-like symbols.

\arccos	\cos	\csc	\exp	\ker	\limsup	\min	\sinh
\arcsin	\cosh	\deg	\gcd	\lg	\ln	\Pr	\sup
\arctan	\cot	\det	\hom	\lim	\log	\sec	\tan
\arg	\coth	\dim	\inf	\liminf	\max	\sin	\tanh

Table B.13. Delimiters.

(	(	)	)
[	[(or \lbrack)	]	] (or \rbrack)
{	\{	}	\}
↑	\uparrow	↓	\downarrow
⇑	\Uparrow	⇓	\Downarrow
↕	\updownarrow	⇕	\Updownarrow
⌊	\lfloor	⌋	\rfloor
⌈	\lceil	⌉	\rceil
⟨	\langle	⟩	\rangle
/	/	\	\backslash
			(or \vert)	‖	| (or \Vert)

Table B.14. Large delimiters.

⎱	\rmoustache	⌠	\lmoustache	⎞	\rgroup
⎛	\lgroup			\arrowvert	‖	\Arrowvert
		\bracevert				

Table B.15. Some other constructions.

$\widetilde{abc}$	\widetilde{abc}	$\widehat{abc}$	\widehat{abc}
$\overleftarrow{abc}$	\overleftarrow{abc}	$\overrightarrow{abc}$	\overrightarrow{abc}
$\overline{abc}$	\overline{abc}	$\underline{abc}$	\underline{abc}
$\overbrace{abc}$	\overbrace{abc}	$\underbrace{abc}$	\underbrace{abc}
$\sqrt{abc}$	\sqrt{abc}	$\sqrt[n]{abc}$	\sqrt[n]{abc}*
f'	f'	$\frac{abc}{xyz}$	\frac{abc}{xyz}*

* Predefined in LaTeX 2_ε but not in TeX.

Table B.16. Spacing.

[]	\!	negative thin space* (normally −1/6 quad)
][	\,	thin space (normally 1/6 of a quad)
][	\:†	medium space* (normally 2/9 of a quad)
] [	\;	thick space* (normally 5/18 of a quad)
] [	\quad	quad
] [	\qquad	qquad (two quads)

* Can appear only in math mode.

† The medium space is \: in LATEX but \> in TEX.

Table B.17. AMS Delimiters.

⌜ \ulcorner	⌝ \urcorner	⌞ \llcorner	⌟ \lrcorner

Table B.18. AMS Arrows.

--→ \dashrightarrow		←-- \dashleftarrow	
⇇ \leftleftarrows		⇆ \leftrightarrows	
⇚ \Lleftarrow		↞ \twoheadleftarrow	
↢ \leftarrowtail		↫ \looparrowleft	
⇋ \leftrightharpoons		↶ \curvearrowleft	
↺ \circlearrowleft		↰ \Lsh	
⇈ \upuparrows		↿ \upharpoonleft	
⇂ \downharpoonleft		⊸ \multimap	
↭ \leftrightsquigarrow		⇉ \rightrightarrows	
⇄ \rightleftarrows		⇉ \rightrightarrows	
⇄ \rightleftarrows		↠ \twoheadrightarrow	
↣ \rightarrowtail		↬ \looparrowright	
⇌ \rightleftharpoons		↷ \curvearrowright	
↻ \circlearrowright		↱ \Rsh	
⇊ \downdownarrows		↾ \upharpoonright	
⇃ \downharpoonright		⇝ \rightsquigarrow	

Table B.19. AMS Negated Arrows.

↚	\nleftarrow	↛	\nrightarrow
⇍	\nLeftarrow	⇏	\nRightarrow
↮	\nleftrightarrow	⇎	\nLeftrightarrow

Table B.20. AMS Greek.

ϝ	\digamma	ϰ	\varkappa

Table B.21. AMS Hebrew.

ℶ	\beth	ℸ	\daleth	ℷ	\gimel

Table B.22. AMS Miscellaneous.

ℏ	\hbar	ℏ	\hslash	△	\vartriangle
▽	\triangledown	□	\square	◇	\lozenge
Ⓢ	\circledS	∠	\angle	∡	\measuredangle
∄	\nexists	℧	\mho	⅃	\Finv
⅁	\Game	ⅈ	\Bbbk	‵	\backprime
∅	\varnothing	▲	\blacktriangle	▼	\blacktriangledown
■	\blacksquare	◆	\blacklozenge	★	\bigstar
◁	\sphericalangle	∁	\complement	ð	\eth
╱	\diagup	╲	\diagdown		

Table B.23. AMS Binary Operators.

∔	\dotplus	∖	\smallsetminus	⋒	\Cap
⋓	\Cup	⊼	\barwedge	⊻	\veebar
⩞	\doublebarwedge	⊟	\boxminus	⊠	\boxtimes
⊡	\boxdot	⊞	\boxplus	⊛	\divideontimes
⋉	\ltimes	⋊	\rtimes	⋋	\leftthreetimes
⋌	\rightthreetimes	⋏	\curlywedge	⋎	\curlyvee
⊖	\circleddash	⊛	\circledast	⊚	\circledcirc
⋅	\centerdot	⊺	\intercal		

Table B.24. AMS Binary Relations.

≦	\leqq	⩽	\leqslant
⪕	\eqslantless	≲	\lesssim
⪅	\lessapprox	≊	\approxeq
⋖	\lessdot	⋘	\lll
≷	\lessgtr	⋚	\lesseqgtr
⪛	\lesseqqgtr	≐	\doteqdot
≓	\risingdotseq	≒	\fallingdotseq
∽	\backsim	⋍	\backsimeq
⊆	\subseteqq	⋐	\Subset
⊏	\sqsubset	≼	\preccurlyeq
⋞	\curlyeqprec	≾	\precsim
⪷	\precapprox	◁	\vartriangleleft
⊴	\trianglelefteq	⊨	\vDash
⊪	\Vvdash	⌣	\smallsmile
⌢	\smallfrown	≏	\bumpeq
≎	\Bumpeq	≧	\geqq
⩾	\geqslant	⪖	\eqslantgtr
≳	\gtrsim	⪆	\gtrapprox
⋗	\gtrdot	⋙	\ggg
≷	\gtrless	⋛	\gtreqless
⪚	\gtreqqless	⋍	\eqcirc
≗	\circeq	⊴	\trianglelq
∼	\thicksim	≈	\thickapprox
⊇	\supseteqq	⋑	\Supset
⊐	\sqsupset	≽	\succcurlyeq
⋟	\curlyeqsucc	≿	\succsim
⪸	\succapprox	▷	\vartriangleright
⊵	\trianglerighteq	⊩	\Vdash
∣	\shortmid	∥	\shortparallel
≬	\between	⋔	\pitchfork
∝	\varpropto	϶	\backepsilon
∴	\therefore	∵	\because
▶	\blacktriangleright	◀	\blacktriangleleft

Table B.25. AMS Negated Binary Relations.

≮	\nless	≰	\nleq	⪇	\nleqslant
≨	\nleqq	⪇	\lneq	⪉	\lneqq
⪇	\lvertneqq	⪉	\lnsim	⪉	\lnapprox
⊀	\nprec	⪯̸	\npreceq	⋨	\precnsim
⪹	\precnapprox	≁	\nsim	∤	\nshortmid
∤	\nmid	⊬	\nvdash	⊭	\nvDash
⋪	\ntriangleleft	⋬	\ntrianglelefteq	⊄	\nsubseteq
⊊	\subsetneq	⊊	\varsubsetneq	⊊	\subsetneqq
⊊	\varsubsetneqq	≯	\ngtr	≱	\ngeq
⪈	\ngeqslant	≩	\ngeqq	⪈	\gneq
⪊	\gneqq	⪈	\gvertneqq	⋩	\gnsim
⪺	\gnapprox	⊁	\nsucc	⪰̸	\nsucceq
⪱	\nsucceq	⋩	\succnsim	⪺	\succnapprox
≇	\ncong	∦	\nshortparallel	∦	\nparallel
⊭	\nvDash	⊯	\nVDash	⋫	\ntriangleright
⋭	\ntrianglerighteq	⊉	\nsupseteq	⊋	\nsupseteqq
⊋	\supsetneq	⊋	\varsupsetneq	⊋	\supsetneqq
⊋	\varsupsetneqq				

Table B.26. Math Alphabets.

		Required package
ABCdef	\mathrm{ABCdef}	
ABCdef	\mathit{ABCdef}	
ABCdef	\mathnormal{ABCdef}	
𝒜ℬ𝒞	\mathcal{ABC}	
𝒜ℬ𝒞	\mathcal{ABC}	euscript with option mathcal
𝔄𝔅ℭ𝔡𝔢𝔣	\mathfrak{ABCdef}	eufrak
𝔸𝔹ℂ	\mathbb{ABC}	amsfonts or amssymb

Appendix C
GNU Emacs—The Sixty+ Most Useful Commands

^x means hold down the control key and press x.

Esc-x means press Esc, release it, then press x (or hold down the meta key and press x).

Use tab key for filename completion in the minibuffer.

General			
^x ^c	exit	^g	panic key—aborts current situation
^x u	undo	^z	suspend Emacs (fg to restart)

Cursor Motion			
^n	next line	^p	previous line
^a	beginning of line	^e	end of line
^f	forward character	^b	back character
Esc-f	forward word	Esc-b	back word
Esc-<	beginning of buffer	Esc->	end of buffer
^v	next page	Esc-v	previous page
^l	redraw screen, centring line		

Deletion			
^d	delete next character	Esc-d	delete next word

Files	
^x ^f	load file
^x i	read file into current buffer
^x ^s	save file
^x s	save all buffers
^x ^w	write named file
Esc-x write-region	save region

Buffers		
^x b switch to another buffer	^x ^b list all buffers
^x ^k kill buffer	

Search and Replace	
^s	incremental search forward*
^r	incremental search backward*
Esc-%	query replace

* terminate with Esc (leave point on found item) or
^g (point remains where it was at start of search).

Cut and Paste (region = from mark to point)	
^k	kill to end of line
^k ^k	kills to end of line then next new line
^w	kill region
Esc-w	copy region to kill ring
^y	yank from kill ring
Esc-y	yank pop (use after ^y)
^@ or ^space	set mark
^x ^x	swap point and mark
^u ^@	goto previous mark
Esc-h	mark paragraph
^x h	mark buffer

Transposing, Capitalizing, Help, etc.			
^t	transpose characters	Esc-t	transpose words
Esc-l	lowercase word	Esc-u	uppercase word
^x ^u	uppercase region	^x ^l	lowercase region
^q	insert literal	^h k	describe key (help)
^h l	show last 100 chars typed	^h t	Emacs tutorial
Esc-q	reformat paragraph	^x ^t	transpose lines

Windows	
^x 2	divide screen into two windows
^x o	switch to other window
^x 1	current window becomes only window

Mail	
^x m	mail buffer
^c ^c	send mail, select other buffer
^c ^s	send mail, leave mail buffer open

Macros	
^x (	start recording keyboard macro
^x)	end recording keyboard macro
^x e	play keyboard macro

Shell	
Esc-!	shell
Esc-|	shell with region as input
Esc-x shell	start shell buffer

Appendix D
Mathematical and Other Organizations

American Mathematical Society
P.O. Box 6248
Providence, Rhode Island 02940-6248
USA
tel: 800-321-4AMS (4267) or 401-455-4000
fax: 401-331-3842
email: ams@math.ams.org
Web: http://www.ams.org/

Canadian Mathematical Society
577 King Edward, Suite 109
POB 450, Station A
Ottawa, Ontario
K1N 6N5, Canada
tel: 613-562-5702
fax: 613-565-1539
email: office@cms.math.ca
Web: http://camel.math.ca/CMS/

Edinburgh Mathematical Society
University of Edinburgh
James Clerk Maxwell Building
Mayfield Road
Edinburgh
EH9 3JZ
Scotland
email: Edmathsoc@maths.ed.ac.uk
Web: http://www.maths.ed.ac.uk/~chris/ems/

239

European Mathematical Society
EMS Secretariat: Mrs. T. Mäkeläinen, Department of Mathematics
P.O. Box 4 (Hallituskatu 15)
FIN-00014 University of Helsinki
Finland
email: tuulikki.makelainen@helsinki.fi
Web: http://www.maths.soton.ac.uk/EMIS/index.html

The Institute of Mathematics and Its Applications
Catherine Richards House
16 Nelson Street
Southend-on-Sea
Essex
SS1 1EF
England
tel: 01702 354020
fax: 01702 354111
email: post@ima.org.uk
Web: http://www.ima.org.uk/

London Mathematical Society
Burlington House
Piccadilly
London
W1V 0NL
England
tel: 0171 437 5377
fax: 0171 439 4629
email: lms@lms.ac.uk
Web: http://www.lms.ac.uk/

Mathematical Association
259 London Road
Leicester
LE2 3BE
England
tel: 0116 2703877
fax: 0116 2703877

Mathematical Association of America
1529 Eighteenth Street, NW
Washington, D.C. 20036-1385
USA
tel: 202-387-5200
fax: 202-265-2384
email: `maahq@maa.org`
Web: `http://www.maa.org`

For member services:
The MAA Service Center
P.O. Box 91112
Washington, D.C. 20090-1112
USA
tel: 800-331-1622, 301-617-7800
fax: 301-206-9789

The Society for Industrial and Applied Mathematics
3600 University City Science Center
Philadelphia, Pennsylvania 19104-2688
USA
tel: 215-382-9800
fax: 215-386-7999
email: `siam@siam.org`
Web: `http://www.siam.org`

TeX Users Group
P.O. Box 1239
Three Rivers, CA 93271-1239
USA
tel: 209-561-0112
fax: 209-561-4584
email: `tug@mail.tug.org`
Web: `http://www.tug.org/`

Appendix E
Winners of Prizes for Expository Writing

This appendix is based on lists supplied by the Mathematical Association of America. The American Mathematical Society also awards prizes for mathematical writing; full details are available on the e-MATH Web page (see §14.1).

Winners of the Chauvenet Prize

Named after William Chauvenet (1820–1870), a professor of mathematics in the United States Navy, this prize is awarded for a "noteworthy paper published in English, such as will come within the range of profitable reading for a member of the Mathematical Association of America." The first twenty-four prize-winning papers (Bliss (1924)–Zalcman (1974)) are collected in the two volume *Chauvenet Papers* [1] (which, usefully, are indexed).

1925

Gilbert Ames Bliss, *Algebraic functions and their divisors*, Ann. Math., 26, 1924, pp. 95–124.

1929

T. H. Hildebrandt, *The Borel theorem and its generalizations*, Bull. Amer. Math. Soc., 32, 1926, pp. 423–474.

1932

G. H. Hardy, *An introduction to the theory of numbers*, Bull. Amer. Math. Soc., 35, 1929, pp. 778–818.

1935

Dunham Jackson, *Series of orthogonal polynomials*, Ann. Math., 2 (34), 1933, pp. 527–545; *Orthogonal trigonometric sums*, Ann. Math., 2 (34),

1933, pp. 799–814; *The convergence of Fourier series*, Amer. Math. Monthly, 1934.

1938

G. T. Whyburn, *On the structure of continua*, Bull. Amer. Math. Soc., 42, 1936, pp. 49–73.

1941

Saunders MacLane, *Modular fields*, Amer. Math. Monthly, 47 (5), 1940, pp. 259–274; *Some recent advances in algebra*, Amer. Math. Monthly, 46 (1), 1939, pp. 3–19.

1944

Robert H. Cameron, *Some introductory exercises in the manipulation of Fourier transforms*, National Mathematics Magazine, 1941, pp. 331–356.

1947

Paul R. Halmos, *The foundations of probability*, Amer. Math. Monthly, 51 (9), 1944, pp. 493–510.

1950

Mark Kac, *Random walk and the theory of Brownian motion*, Amer. Math. Monthly, 54 (7), 1947, pp. 369–391.

1953

E. J. McShane, *Partial orderings and Moore–Smith limits*, Amer. Math. Monthly, 59 (1), 1952, pp. 1–11.

1956

R. H. Bruck, *Recent advances in the foundations of Euclidean plane geometry*, Amer. Math. Monthly, 52 (7, Part II), 1955, pp. 2–17.

1960

Cornelius Lanczos, *Linear systems in self-adjoint form*, Amer. Math. Monthly, 65 (9), 1958, pp. 665–679.

1963

Philip J. Davies, *Leonhard Euler's integral: A historical profile of the gamma function*, Amer. Math. Monthly, 66 (10), 1959, pp. 849–869.

1964

Leon A. Henkin, *Are logic and mathematics identical?*, Science, 138 (3542), 1962, pp. 788–794.

1965

Jack K. Hale and Joseph P. LaSalle, *Differential equations: Linearity vs. nonlinearity*, SIAM Rev., 5 (3), 1963, pp. 249–272.

1966

No award

1967

Guido L. Weiss, *Harmonic analysis*, MAA Stud. Math., 3, 1965, pp. 124–178.

1968

Mark Kac, *Can one hear the shape of a drum?*, Amer. Math. Monthly, 73 (4, Part II), Slaught Papers No. 11, 1966, pp. 1–23.

1969

No award

1970

Shiing-Shen Chern, *Curves and surfaces in Euclidean space*, MAA Stud. Math., 1967, pp. 16–56.

1971

Norman Levinson, *A motivated account of an elementary proof of the prime number theorem*, Amer. Math. Monthly, 76 (2), 1969, pp. 225–245.

1972

Jean François Trèves, *On local solvability of linear partial differential equations*, Bull. Amer. Math. Soc., 76, 1970, pp. 552–571.

1973

Carl D. Olds, *The simple continued fraction expansion of e*, Amer. Math. Monthly, 77 (9), 1970, pp. 968–974.

1974

Peter D. Lax, *The formation and decay of shock waves*, Amer. Math. Monthly, 79 (3), 1972, pp. 227–241.

1975

Martin D. Davis and Reuben Hersh, *Hilbert's 10th problem*, Scientific American, 229 (5), November 1973, pp. 84–91.

1976

Lawrence Zalcman, *Real proofs of complex theorems (and vice versa)*, Amer. Math. Monthly, 81 (2), 1974, pp. 115–137.

1977

W. Gilbert Strang, *Piecewise polynomials and the finite element method*, Bull. Amer. Math. Soc., 79 (6), 1973, 1128–1137.

1978

Shreeram S. Abhyankar, *Historical ramblings in algebraic geometry and related algebra*, Amer. Math. Monthly, 83 (6), 1976, pp. 409–448.

1979

Neil J. A. Sloane, *Error-correcting codes and invariant theory: New applications of a nineteenth-century technique*, Amer. Math. Monthly, 84 (2), 1977, pp. 82–107.

1980

Heinz Bauer, *Approximation and abstract boundaries*, Amer. Math. Monthly, 85 (8), 1978, pp. 632–647.

1981

Kenneth I. Gross, *On the evolution of noncommutative harmonic analysis*, Amer. Math. Monthly, 85 (7), 1978, pp. 525–548.

1982

No award

1983

No award

1984

R. Arthur Knoebel, *Exponentials reiterated*, Amer. Math. Monthly, 88 (4), 1981, pp. 235–252.

1985

Carl Pomerance, *Recent developments in primality testing*, Mathematical Intelligencer, 3 (3), 1981, pp. 97–105.

1986

George Miel, *Of calculations past and present: The Archimedean algorithm*, Amer. Math. Monthly, 90 (17), 1983, pp. 17–35.

1987

James H. Wilkinson, *The perfidious polynomial*, in Gene H. Golub, ed., Studies in Numerical Analysis, vol. 24 of Studies in Mathematics, The Mathematical Association of America, Washington, D.C., 1984, pp. 1–28.

1988

Steve Smale, *On the efficiency of algorithms of analysis*, Bull. Amer. Math. Soc., 13 (2), 1985, pp. 87–121.

1989

Jacob Korevaar, *Ludwig Bieberbach's conjecture and its proof by Louis de Branges*, Amer. Math. Monthly, 93 (7), 1986, pp. 505–514.

1990

David Allen Hoffman, *The computer-aided discovery of new embedded minimal surfaces*, Mathematical Intelligencer, 9, 1987.

1991

W. B. R. Lickorish and Kenneth C. Millett, *The new polynomial invariants of knots and links*, Math. Mag., 61 (1), 1988, pp. 3–23.

1992

Steven G. Krantz, *What is several complex variables?*, Amer. Math. Monthly, 94 (3), 1987, pp. 236–256.

1993

J. M. Borwein, P. B. Borwein, and D. H. Bailey, *Ramanujan, modular equations, and approximations to Pi or how to compute one billion digits of Pi*, Amer. Math. Monthly, 96 (3), 1989, pp. 201–219.

1994

Barry Mazur, *Number theory as gadfly*, Amer. Math. Monthly, 98 (7), 1991, pp. 593–610.

1995

Donald G. Saari, *A visit to the Newtonian N-body problem via elementary complex variables*, Amer. Math. Monthly, 97 (2), 1990, pp. 105–119.

1996

Joan Birman, *New points of view in knot theory*, Bull. Amer. Math. Soc., 28, 1993, pp. 253–287.

1997

Tom Hawkins, *The birth of Lie's theory of groups*, The Mathematical Intelligencer, 1994, pp. 6–17.

Winners of the Lester R. Ford Award

An award for articles in the American Mathematical Monthly.

1965

R. H. Bing, *Spheres in E^3*, Amer. Math. Monthly, 71 (4), 1964, pp. 353–364.

Louis Brand, *A division algebra for sequences and its associated operational calculus*, Amer. Math. Monthly, 71 (7), 1964, pp. 719–728.

R. G. Kuller, *Coin tossing, probability, and the Weierstrass approximation theorem*, Math. Mag., 37, 1964, pp. 262–265.

R. D. Luce, *The mathematics used in mathematical psychology*, Amer. Math. Monthly, 71 (4), 1964, pp. 364–378.

Hartley Rogers, Jr., *Information theory*, Math. Mag., 37, 1964, pp. 63–78.

Elmer Tolsted, *An elementary derivation of the Cauchy, Hölder, and Minkowski inequalities from Young's inequality*, Math. Mag., 37, 1964, pp. 2–12.

1966

C. B. Allendoerfer, *Generalizations of theorems about triangles*, Math. Mag., 38, 1965, pp. 253–259.

Peter D. Lax, *Numerical solution of partial differential equations*, Amer. Math. Monthly, 72 (2, Part II), Slaught Papers No. 10, 1965, pp. 74–84.

Marvin Marcus and Henryk Minc, *Permanents*, Amer. Math. Monthly, 72 (6), 1965, pp. 577–591.

1967

Wai-Kai Chen, *Boolean matrices and switching nets*, Math. Mag., 36, 1966, pp. 1–8.

D. R. Fulkerson, *Flow networks and combinatorial operations research*, Amer. Math. Monthly, 73 (2), 1966, pp. 115–138.

Mark Kac, *Can one hear the shape of a drum?*, Amer. Math. Monthly, 73 (4, Part II), Slaught Papers No. 11, 1966, pp. 1–23.

M. Z. Nashed, *Some remarks on variations and differentials*, Amer. Math. Monthly, 73 (4, Part II), Slaught Papers No. 11, 1966, pp. 63–76.

P. B. Yale, *Automorphisms of the complex numbers*, Math. Mag., 39, 1966, pp. 135–141.

1968

Frederic Cunningham, Jr., *Taking limits under the integral sign*, Math. Mag., 40, 1967, pp. 179–186.

W. F. Newns, *Functional dependence*, Amer. Math. Monthly, 74 (8), 1967, pp. 911–920.

Daniel Pedoe, *On a theorem in geometry*, Amer. Math. Monthly, 74 (6), 1967, pp. 627–640.

Keith L. Phillips, *The maximal theorems of Hardy and Littlewood*, Amer. Math. Monthly, 74 (6), 1967, pp. 648–660.

F. V. Waugh and Margaret W. Maxfield, *Side-and-diagonal numbers*, Math. Mag., 40, 1967, pp. 74–83.

Hans J. Zassenhaus, *On the fundamental theorem of algebra*, Amer. Math. Monthly, 74 (5), 1967, pp. 485–497.

1969

Harley Flanders, *A proof of Minkowski's inequality for convex curves*, Amer. Math. Monthly, 75 (6), 1968, pp. 581–593.

George E. Forsythe, *What to do till the computer scientist comes*, Amer. Math. Monthly, 75 (5), 1968, pp. 454–462.

M. F. Neuts, *Are many 1-1-functions on the positive integers onto?*, Math. Mag., 41, 1968, pp. 103–109.

Pierre Samuel, *Unique factorization*, Amer. Math. Monthly, 75 (9), 1968, pp. 945–952.

Hassler Whitney, *The mathematics of physical quantities*, I *and* II, Amer. Math. Monthly, 75 (2,3), 1968, pp. 115–138 and 227–256.

Albert Wilansky, *Spectral decomposition of matrices for high school students*, Math. Mag., 41, 1968, pp. 51–59.

1970

Henry L. Alder, *Partition identities—from Euler to the present*, Amer. Math. Monthly, 76 (7), 1969, pp. 733–746.

Ralph P. Boas, *Inequalities for the derivatives of polynomials*, Math. Mag., 42, 1969, pp. 165–174.

W. A. Coppel, *J. B. Fourier—on the occasion of his two hundredth birthday*, Amer. Math. Monthly, 76 (5), 1969, pp. 468–483.

Norman Levinson, *A motivated account of an elementary proof of the prime number theorem*, Amer. Math. Monthly, 3, 1969, pp. 225–245.

John Milnor, *A problem in cartography*, Amer. Math. Monthly, 10, 1969, pp. 1101–1112.

Ivan Niven, *Formal power series*, Amer. Math. Monthly, 8, 1969, pp. 871–889.

1971

Jean A. Dieudonné, *The work of Nicholas Bourbaki*, Amer. Math. Monthly, 77 (2), 1970, pp. 134–145.

George E. Forsythe, *Pitfalls in computation, or why a math book isn't enough*, Amer. Math. Monthly, 77 (9), 1970, pp. 931–956.

Paul R. Halmos, *Finite-dimensional Hilbert spaces*, Amer. Math. Monthly, 77 (5), 1970, pp. 457–464.

Eric Langford, *A problem in geometric probability*, Math. Mag., 43, 1970, pp. 237–244.

P. V. O'Neil, *Ulam's conjecture and graph reconstructions*, Amer. Math. Monthly, 77 (1), 1970, pp. 35–43.

Olga Taussky, *Sums of squares*, Amer. Math. Monthly, 77 (8), 1970, pp. 805–830.

1972

G. D. Chakerian and L. H. Lange, *Geometric extremum problems*, Math. Mag., 44, 1971, pp. 57–69.

P. M. Cohn, *Rings of fractions*, Amer. Math. Monthly, 78 (6), 1971, pp. 596–615.

Frederic Cunningham, Jr., *The Kakeya problem for simply connected and for star-shaped sets*, Amer. Math. Monthly, 78 (2), 1971, pp. 114–129.

W. J. Ellison, *Waring's problem*, Amer. Math. Monthly, 78 (1), 1971, pp. 10–36.

Leon Henkin, *Mathematical foundations for mathematics*, Amer. Math. Monthly, 78 (5), 1971, pp. 463–487.

Victor Klee, *What is a convex set?*, Amer. Math. Monthly, 78 (6), 1971, pp. 616–631.

1973

Jean A. Dieudonné, *The historical development of algebraic geometry*, Amer. Math. Monthly, 79 (8), 1972, pp. 827–866.

Samuel Karlin, *Some mathematical models of population genetics*, Amer. Math. Monthly, 79 (7), 1972, pp. 699–739.

Peter D. Lax, *The formation and decay of shock waves*, Amer. Math. Monthly, 79 (3), 1972, pp. 227–241.

Thomas L. Saaty, *Thirteen colorful variations on Guthrie's four-color conjecture*, Amer. Math. Monthly, 79 (1), 1972, pp. 2–43.

Lynn A. Steen, *Conjectures and counterexamples in metrization theory*, Amer. Math. Monthly, 79 (2), 1972, pp. 113–132.

R. L. Wilder, *History in the mathematics curriculum: Its status, quality, and function*, Amer. Math. Monthly, 79 (5), 1972, pp. 479–495

1974

Patrick Billinglsey, *Prime numbers and Brownian motion*, Amer. Math. Monthly, 80 (10), 1973, pp. 1099–1115.

Garrett Birkhoff, *Current trends in algebra*, Amer. Math. Monthly, 80 (7), 1973, pp. 760–782.

Martin Davis, *Hilbert's tenth problem is unsolvable*, Amer. Math. Monthly, 80 (3), 1973, pp. 233–269.

I. J. Schoenberg, *The elementary cases of Landau's problem of inequalities between derivatives*, Amer. Math. Monthly, 80 (2), 1973, pp. 121–158.

Lynn A. Steen, *Highlights in the history of spectral theory*, Amer. Math. Monthly, 80 (4), 1973, pp. 359–381

Robin J. Wilson, *An introduction to matroid theory*, Amer. Math. Monthly, 80 (5), 1973, pp. 500–525.

1975

Raymond Ayoub, *Euler and the zeta function*, Amer. Math. Monthly, 81 (10), 1974, pp. 1067–1086.

James Callahan, *Singularities and plane maps*, Amer. Math. Monthly, 81 (3), 1974, pp. 211–240.

Donald E. Knuth, *Computer science and its relation to mathematics*, Amer. Math. Monthly, 81 (4), 1974, pp. 323–343.

Johannes C. C. Nitsche, *Plateau's problems and their modern ramifications*, Amer. Math. Monthly, 81 (9), 1974, pp. 945–968.

S. K. Stein, *Algebraic tiling*, Amer. Math. Monthly, 81 (5), 1974, pp. 445–462.

Lawrence Zalcman, *Real proofs of complex theorems (and vice versa)*, Amer. Math. Monthly, 81 (2), 1974, pp. 115–137.

1976

M. L. Balinski and H. P. Young, *The quota method of apportionment*, Amer. Math. Monthly, 82 (7), 1975, pp. 701–730.

E. A. Bender and J. R. Goldman, *On the applications of Mobius inversion in combinatorial analysis*, Amer. Math. Monthly, 82 (8), 1975, pp. 789–803.

Branko Grünbaum, *Venn diagrams and independent families of sets*, Math. Mag., 48, 1975, pp. 12–23.

J. E. Humphreys, *Representations of $SL(2, p)$*, Amer. Math. Monthly, 82 (1), 1975, pp. 21–39.

J. B. Keller and D. W. McLaughlin, *The Feynman integral*, Amer. Math. Monthly, 82 (5), 1975, pp. 451–465.

J. J. Price, *Topics in orthogonal functions*, Amer. Math. Monthly, 82 (6), 1975, pp. 594–609.

1977

Shreeram S. Abhyankar, *Historical ramblings in algebraic geometry and related algebra*, Amer. Math. Monthly, 83 (6), 1976, pp. 409–448.

Joseph B. Keller, *Inverse problems*, Amer. Math. Monthly, 83 (2), 1976, pp. 107–118.

D. S. Passman, *What is a group ring?*, Amer. Math. Monthly, 83 (3), 1976, pp. 173–185.

James P. Jones, Diahachiro Sato, Hideo Wada, and Douglas Wiens, *Diophantine representation of the set of prime numbers*, Amer. Math. Monthly, 83 (6), 1976, pp. 449–464.

J. H. Ewing, W. H. Gustafson, P. R. Halmos, S. H. Moolgavkar, W. H. Wheeler, and W. P. Ziemer, *American mathematics from 1940 to the day before yesterday*, Amer. Math. Monthly, 83 (7), 1976, pp. 503–516.

1978

Ralph P. Boas, Jr., *Partial sums of infinite series, and how they grow*, Amer. Math. Monthly, 84 (4), 1977, pp. 237–258.

Louis H. Kauffman and Thomas F. Banchoff, *Immersions and mod-2 quadratic forms*, Amer. Math. Monthly, 84 (3), 1977, pp. 168–185.

Neil J. A. Sloane, *Error-correcting codes and invariant theory: New applications of a nineteenth-century technique*, Amer. Math. Monthly, 84 (2), 1977, pp. 82–107.

1979

Bradley Efron, *Controversies in the foundations of statistics*, Amer. Math. Monthly, 85 (4), 1978, pp. 231–246.

Ned Glick, *Breaking records and breaking boards*, Amer. Math. Monthly, 85 (1), 1978, pp. 2–26.

Kenneth I. Gross, *On the evolution of noncommutative harmonic analysis*, Amer. Math. Monthly, 85 (7), 1978, pp. 525–548.

Lawrence A. Shepp and Joseph B. Kruskal, *Computerized tomography: The new medical X-ray technology*, Amer. Math. Monthly, 85 (6), 1978, pp. 420–439.

1980

Desmond P. Fearnley-Sander, *Hermann Grassmann and the creation of linear algebra*, Amer. Math. Monthly, 86 (10), 1979, pp. 809–817.

David Gale, *The game of hex and the Brouwer fixed-point theorem*, Amer. Math. Monthly, 86 (10), 1979, pp. 818–826.

Karel Hrbacek, *Nonstandard set theory*, Amer. Math. Monthly, 86 (8), 1979, pp. 659–677.

Cathleen S. Morawetz, *Nonlinear conservation equations*, Amer. Math. Monthly, 86 (4), 1979, pp. 284–287.

Robert Osserman, *Bonnesen-style isoperimetric inequalities*, Amer. Math. Monthly, 86 (1), 1979, pp. 1–29.

1981

R. Creighton Buck, *Sherlock Holmes in Babylon*, Amer. Math. Monthly, 87 (5), 1980, pp. 335–345.

Brune H. Pourciau, *Modern multiplier rules*, Amer. Math. Monthly, 87 (6), 1980, pp. 433-452.

Alan H. Schoenfeld, *Teaching problem-solving skills*, Amer. Math. Monthly, 87 (10), 1980, pp. 794–805.

Edward R. Swart, *The philosophical implications of the four-color problem*, Amer. Math. Monthly, 87 (9), 1980, pp. 697–707.

Lawrence A. Zalcman, *Offbeat integral geometry*, Amer. Math. Monthly, 87 (3), 1980, pp. 161–175.

1982

Philip J. Davis, *Are there coincidences in mathematics?*, Amer. Math. Monthly, 88 (5), 1981, pp. 311–320.

R. Arthur Knoebel, *Exponentials reiterated*, Amer. Math. Monthly, 88 (4), 1981, pp. 235–252.

1983

Robert F. Brown, *The fixed point property and Cartesian products*, Amer. Math. Monthly, 89 (9), 1982, pp. 654–678.

Tony Rothman, *Genius and biographers: The fictionalization of Evariste Galois*, Amer. Math. Monthly, 89 (2), 1982, pp. 84–106.

Robert S. Strichartz, *Radon inversion—variations on a theme*, Amer. Math. Monthly, 89 (6), 1982, pp. 377–384 and 420–423 (solutions of problems).

1984

Judith V. Grabiner, *Who gave you the epsilon? Cauchy and the origins of rigorous calculus*, Amer. Math. Monthly, 90 (3), 1983, pp. 185–194.

Roger Howe, *Very basic Lie theory*, Amer. Math. Monthly, 90 (9), 1983, pp. 600–623.

John Milnor, *On the geometry of the Kepler problem*, Amer. Math. Monthly, 90 (6), 1983, pp. 353–365.

Joel Spencer, *Large numbers and unprovable theorems*, Amer. Math Monthly, 90 (10), 1983, pp. 669–675.

William C. Waterhouse, *Do symmetric problems have symmetric solutions?*, Amer. Math. Monthly, 90 (6), 1983, pp. 378–387.

1985

John D. Dixon, *Factorization and primality tests*, Amer. Math. Monthly, 91 (6), 1984, pp. 333–352.

Donald G. Saari and John B. Urenko, *Newton's method, circle maps, and chaotic motion*, Amer. Math. Monthly, 91 (1), 1984, pp. 3–17.

1986

Jeffrey C. Lagarias, *The $3x+1$ problem and its generalizations*, Amer. Math. Monthly, 92 (1), 1985, pp. 3–23.

Michael E. Taylor, *Review of Lars Hörmander's "The Analysis of linear partial differential operations, I and II"*, Amer. Math. Monthly, 92 (10), 1985, pp. 745–749.

1987

Stuart S. Antman, *Review of Ann Hibler Koblitz's "A convergence of lives— Sophia Kovalevskaia: Scientist, Writer, Revolutionary"*, Amer. Math. Monthly, 93 (2), 1986, pp. 139–144.

Joan Cleary, Sidney A. Morris and David Yost, *Numerical geometry— numbers for shapes*, Amer. Math. Monthly, 93 (4), 1986, pp. 260–275.

Howard Hiller, *Crystallography and cohomology of groups*, Amer. Math. Monthly, 93 (10), 1986, pp. 765–779.

Jacob Korevaar, *Ludwig Bieberbach's conjecture and its proof by Louis de Branges*, Amer. Math. Monthly, 93 (7), 1986, pp. 505–514.

Peter M. Neumann, *Review of Harold M. Edwards' "Galois Theory"*, Amer. Math. Monthly, 93 (5), 1986, pp. 407–411.

1988

James F. Epperson, *On the Runge example*, Amer. Math. Monthly, 94 (4), 1987, pp. 329–341.

Stan Wagon, *Fourteen proofs of a result about tiling a rectangle*, Amer. Math. Monthly, 94 (7), 1987, pp. 601–617.

1989

Richard K. Guy, *The strong law of small numbers*, Amer. Math. Monthly, 95 (8), 198, pp. 697–712.

Gert Almkvist and Bruce Berndt, *Gauss, Landen, Ramanujan, the arithmetic-geometric mean, ellipses, π and the "Ladies Diary"*, Amer. Math. Monthly, 95 (7), 1988, pp. 585–608.

1990

Jacob E. Goodman, János Pach and Chee K. Yap, *Mountain climbing, ladder moving and the ring-width of a polygon*, Amer. Math. Monthly, 96 (6), 1989, pp. 494–510.

Doron Zeilberger, *Kathy O'Hara's constructive proof of the unimodality of the Gaussian polynomials*, Amer. Math. Monthly, 96 (7), 1989, pp. 590–602.

1991

Marcel Berger, *Convexity*, Amer. Math. Monthly, 97 (8), 1990, pp. 650–678.

Ronald L. Graham and Frances Yao, *A whirlwind tour of computational geometry*, Amer. Math. Monthly, 97 (8), 1990, pp. 687–701.

Joyce Justicz, Edward R. Scheinerman and Peter M. Winkler, *Random intervals*, Amer. Math. Monthly, 97 (10), 1990, pp. 881–889.

1992

Clement W. H. Lam, *The search for a finite projective plane of order 10*, Amer. Math. Monthly, 98, 1991, pp. 305–318.

1994

Bruce C. Berndt and S. Bhargava, *Ramanujan—For lowbrows*, Amer. Math. Monthly, 100, 1993, pp. 644–656.

Reuben Hersh, *Szeged in* 1934, Amer. Math. Monthly, 100, 1993, pp. 219–230.

Leonard Gillman, *An axiomatic approach to the integral*, Amer. Math. Monthly, 100, 1993, pp. 16–25.

Joseph H. Silverman, *Taxicabs and sums of two cubes*, Amer. Math. Monthly, 100, 1993, pp. 331–340.

Dan Velleman and István Szalkai, *Versatile coins*, Amer. Math. Monthly, 100, 1993, pp. 26-33.

1995

Fernando Q. Gouvea, *A marvelous proof*, Amer. Math. Monthly, 101, 1994, pp. 203–222.

Jonathan L. King, *Three problems in search of a measure*, Amer. Math. Monthly, 101, 1994, pp. 609–628.

I. Kleiner and N. Movshovitz-Hadar, *The role of paradoxes in the evolution of mathematics*, Amer. Math. Monthly, 102, 1994, pp. 963–974.

William C. Waterhouse, *A counterexample for Germain*, Amer. Math. Monthly, 101, 1994, pp. 140–150.

1996

Martin Aigner, *Turan's graph theorem*, Amer. Math. Monthly, 102, 1995, pp. 808–816.

Sheldon Axler, *Down with determinants!*, Amer. Math. Monthly, 102, 1995, pp. 139-154.

John Oprea, *Geometry and the Foucault pendulum*, Amer. Math. Monthly, 102, 1995, pp. 515–522.

1997

Robert G. Bartle, *Return to the Riemann integral*, Amer. Math. Monthly, 103, 1996, pp. 625–632.

A. F. Beardon, *Sums of powers of integers*, Amer. Math. Monthly, 103, 1996, pp. 201–213.

John Brillhart and Patrick Morton, *A case study in mathematical research: The Golay-Rudin-Shapiro sequence*, Amer. Math. Monthly, 103, 1996, pp. 854–869.

Winners of the George Polya Award

An award for articles in the College Mathematics Journal—formerly the Two-Year College Mathematics Journal.

1977

Julian Weisglass, *Small groups: An alternative to the lecture method*, 7 (1), 1976, pp. 15–20.

Anneli Lax, *Linear algebra, A potent tool*, 7 (2), 1976, pp. 3–15.

1978

Allen H. Holmes, Walter J. Sanders and John W. LeDuc, *Statistical inference for the general education student—it can be done*, 8 (4), 1977, pp. 223–230.

Freida Zames, *Surface area and the cylinder area paradox*, 8 (4), 1977, pp. 207–211.

1979

Richard L. Francis, *A note on angle construction*, 9 (2), 1978, pp. 75–80.

Richard Plagge, *Fractions without quotients: Arithmetic of repeating decimals*, 9 (1), 1978, pp. 11–15.

1980

Hugh F. Ouellette and Gordon Bennett, *The discovery of a generalization*, 10 (2), 1979, pp. 100–106.

Robert Nelson, *Pictures, probability and paradox*, 10 (3), 1979, pp. 182–190.

1981

Gulbank D. Chakerian, *Circles and spheres*, 11 (1), 1980, pp. 26–41.

Dennis D. McCune, Robert G. Dean and William D. Clark, *Calculators to motivate infinite composition of functions*, 11 (3), 1980, pp. 189–195.

1982

John A. Mitchem, *On the history and solution of the four-color problem*, 12 (2), 1981, pp. 108–116.

Peter L. Renz, *Mathematical proof: What it is and what it ought to be*, 12 (2), 1981, pp. 83–103.

1983

Douglas R. Hofstadter, *Analogies and metaphors to explain Gödel's theorem*, 13 (2), 1982, pp. 98–114.

Paul R. Halmos, *The thrills of abstraction*, 13 (4), September 1982, pp. 243–251.

Warren Page and V. N. Murty, *Nearness relations among measures of central tendency and dispersion, Part* 1, 13 (5), 1982, pp. 315–327.

1984

Ruma Falk and Maya Bar-Hillel, *Probabilistic dependence between events,* 14 (3), 1983, pp. 240–247.

Richard J. Trudeau, *How big is a point?,* 14 (4), 1983, pp. 295–300.

1985

Anthony Barcellos, *The fractal geometry of Mandelbrot,* 15 (2), 1984, pp. 98–114.

Kay W. Dundas, *To build a better box,* 15 (1), 1984, pp. 30–36.

1986

Philip J. Davis, *What do I know? A study of mathematical self-awareness,* 16 (1), 1985, pp. 22–41.

1987

Constance Reid, *The autobiography of Julia Robinson,* 17 (1), 1986, pp. 3–21.

Irl C. Bivens, *What a tangent line is when it isn't a limit,* 17 (2), 1986, pp. 133–143.

1988

Dennis M. Luciano and Gordon M. Prichett, *From Caesar ciphers to public-key cryptosystems,* 18 (1), 1987, pp. 2–17.

V. Frederick Rickey, *Isaac Newton: Man, myth and mathematics,* 18 (5), 1987, pp. 362–389.

1989

Edward Rozema, *Why should we pivot in Gaussian elimination?,* 19 (1), 1988, pp. 63–72.

Beverly L. Brechner and John C. Mayer, *Antoine's necklace or how to keep a necklace from falling apart,* 19 (4), 1988, pp. 306–320.

1990

Israel Kleiner, *Evolution of the function concept: A brief survey,* 20 (4), 1989, pp. 282–300.

Richard D. Neidinger, *Automatic differentiation and APL,* 20 (3), 1989, pp. 238–251.

1991

William B. Gearhart and Harris S. Shultz, *The function* $\sin x/x$, 21 (2), 1990, pp. 90–99.

Mark Schilling, *The longest run of heads*, 21 (3), 1990, pp. 196–207.

1993

William Dunham, *Euler and the Fundamental Theorem of Algebra*, 22 (4), 1991.

Howard Eves, *Two surprising theorems on Cavalieri Congruence*, 22 (2), 1991.

1994

Charles W. Groetsch, *Inverse problems and Torricelli's law*, 24 (3), 1993, pp. 210–217.

Dan Kalman, *Six ways to sum a series*, 24 (5), 1993, pp. 402–421.

1995

Anthony P. Ferzola, *Euler and differentials*, 25 (2), 1994.

Paulo Ribenboim, *Prime number records*, 25 (4), 1994.

1996

John H. Ewing, *Can we see the Mandelbrot set?*, 26 (2), 1995, pp. 90–99.

James G. Simmonds, *A new look at an old function, $e^{i\theta}$*, 26 (1), 1995, pp. 6–10.

1997

Chris Christensen, *Newton's method for resolving affected equations*, 27 (5), 1996, pp. 330–340.

Leon Harkleroad, *How mathematicians know what computers can't do*, 27 (1), 1996, pp. 37–42.

Winners of the Carl B. Allendoerfer Award

An award for articles in Mathematics Magazine.

1977

Joseph A. Gallian, *The search for finite groups*, 49 (4), 1976, pp. 163–179.

B. L. van der Waerden, *Hamilton's discovery of quaternions*, 49 (5), 1976, pp. 227–234.

1978

Geoffrey C. Shephard and Branko Grünbaum, *Tilings by regular polygons*, 50 (5), 1977, pp. 227–247.

David A. Smith, *Human population growth: Stability or explosion*, 50 (4), 1977, pp. 186–197.

1979

Doris W. Scattschneider, *Tiling the plane with congruent pentagons*, 51 (1), 1978, pp. 29–44.

Bruce C. Berndt, *Ramanujan's Notebooks*, 51 (3), 1978, pp. 147–164.

1980

Ernst Snapper, *The three crises in mathematics: Logicism, intuitionism, and formalism*, 52 (4), 1979, pp. 207–216.

Victor L. Klee, Jr., *Some unsolved problems in plane geometry*, 52 (3), 1979, pp. 131–145.

1981

Stephen B. Maurer, *The king chicken theorems*, 53 (2), 1980, pp. 67–80.

Donald E. Sanderson, *Advanced plane topology from an elementary standpoint*, 53 (2), 1980, pp. 81–89.

1982

J. Ian Richards, *Continued fractions without tears*, 54 (4), 1981, pp. 163–171.

Marjorie Senechal, *Which tetrahedra fill space?*, 54 (5), 1981, pp. 227–243.

1983

Donald O. Koehler, *Mathematics and literature*, 55 (2), 1982, pp. 81–95.

Clifford H. Wagner, *A generic approach to iterative methods*, 55 (5), 1982, pp. 259–273.

1984

Judith Grabiner, *The changing concept of change: The derivative from Fermat to Weierstrass*, 56 (4), 1983, pp. 195–206.

1985

Philip D. Straffin, Jr. and Bernard Grofman, *Parliamentary coalitions: A tour of models*, 57 (5), 1984, pp. 259–274.

Frederick S. Gass, *Constructive ordinal notation systems*, 57 (3), 1984, pp. 131–141.

1986

Bart Braden, *The design of an oscillating sprinkler*, 58 (1), 1985, pp. 29–38.

Saul Stahl, *The other map coloring theorem*, 58 (3), 1985, pp. 131–145.

1987

Israel Kleiner, *The evolution of group theory*, 59 (4), 1986, pp. 195–215.

Paul Zorn, *The Bieberbach conjecture*, 59 (3), 1986, pp. 131–148.

1988

Steven Galovich, *Products of sines and cosines*, 60 (2), 1987, pp. 105–113.

Bart Braden, *Polya's geometric picture of complex contour integrals*, 60 (5), 1987, pp. 321–327.

1989

Judith V. Grabiner, *The centrality of mathematics in the history of western thought*, 61 (4), 1988, pp. 220–230.

W. B. Raymond Lickorish and Kenneth C. Millett, *The new polynomial invariants of knots and links*, 61 (1), 1988, pp. 3–23.

1990

Fan K. Chung, Martin Gardner, and Ronald L. Graham, *Steiner trees on a checkerboard*, 62 (2), 1989, pp. 83–96.

Thomas Archibald, *Connectivity and smoke-rings: Green's second identity in its first fifty years*, 62 (4), 1989, pp. 219–237.

1991

Ranjan Roy, *The discovery of the series formula for π by Leibniz, Gregory and Nilakantha*, 63 (5), 1990, pp. 291–306.

1992

Israel Kleiner, *Rigor and proof in mathematics: A historical perspective*, 64 (5), 1991, pp. 291–314.

G. D. Chakerian and David Logothetti, *Cube slices, pictorial triangles, and probability*, 64 (4), 1991, pp. 219–241.

1994

Joan Hutchinson, *Coloring ordinary maps, maps of empires, and maps of the moon*, 66 (4), 1993, pp. 211–226.

1995

Lee Badger, *Lazzarini's lucky approximation of π*, 67 (2), 1994.

Tristan Needham, *The geometry of harmonic functions*, 67 (2), 1994.

1996

Judith Grabiner, *Descartes and problem-solving*, 1995, pp. 83–97.

Daniel J. Velleman and Gregory S. Call, *Permutations and combination locks*, 68, 1995, pp. 243–253.

1997

Colm Mulcahy, *Plotting and scheming with wavelets*, 69, December 1996, pp. 323–343.

Lin Tan, *Group of rational points on the unit circle*, 69, June 1996.

Winners of the MAA Book Prize, renamed Beckenbach Book Prize (1986)

1984

Charles Robert Hadlock, *Field Theory and Its Classical Problems*, Carus Monograph No. 19, 1978.

1986

Edward Packel, *The Mathematics of Games and Gambling*, MAA New Mathematical Library Series, 1981.

1989

Thomas M. Thompson, *From Error-Correcting Codes through Sphere Packings to Simple Groups*, Carus Monograph No. 21, 1984.

1994

Steven George Krantz, *Complex Analysis: The Geometric Viewpoint*, Carus Mathematical Monographs, 1990.

1996

Constance Reid, *The Search for E. T. Bell, Also Known as John Taine*, Mathematical Association of America, 1993, x+372 pp. ISBN 0-88385-508-9.

Winners of the Merten M. Hasse Prize (established 1986)

1987

Anthony Barcellos, *The fractal geometry of Mandelbrot*, The College Mathematics Journal, 15, 1984.

1989

Irl C. Bivens, *What a tangent is when it isn't a limit*, The College Mathematics Journal, 17, 1986.

1991

Barry A. Cipra, *An introduction to the Ising model*, Amer. Math. Monthly, 94 (10), 1987, pp. 937–959.

1993

J. M. Borwein, P. B. Borwein, and D. H. Bailey, *Ramanujan, modular equations, and approximations to Pi or how to compute one billion digits of Pi*, Amer. Math. Monthly, 96 (3), 1989, pp. 201–219.

1995

Andrew J. Granville, *Zaphod Beeblebrox's brain and the fifty-ninth row of Pascal's triangle*, Amer. Math. Monthly, 99, 1992, pp. 318–331.

1997

Jonathan King, *Three problems in search of a measure*, Amer. Math. Monthly, 101, August–September 1994.

Glossary

abstract. A brief, self-contained summary of the contents of a paper that appears by itself at the beginning of the paper. Also a brief (written) summary of the contents of a talk.

ACM. The Association for Computing Machinery.

acknowledgements. A section preceding the references (or a footnote, or a final paragraph) in which the author thanks people or organizations for help, advice or financial support for the work described.

AMS. The American Mathematical Society.

$\mathcal{A}_{\mathcal{M}}S$-TEX. A macro package for TEX that makes it easier to typeset mathematical papers with TEX. It gives new structures for displaying mathematical equations and comes with a special font of mathematical symbols.

$\mathcal{A}_{\mathcal{M}}S$-LATEX. A package for LATEX that incorporates the features of $\mathcal{A}_{\mathcal{M}}S$-TEX.

anonymous ftp. A form of ftp in which the user logs on as user `anonymous` and need not type a password (though, by convention, the user's electronic mail address is typed as the password).

archive. In Unix, a single file that contains a set of other files (e.g., as manipulated with the `tar` command). Or a collection of software that is located at a particular Internet address and can be accessed by anonymous ftp.

ASCII. American Standard Code for Information Interchange. A coding system in which letters, digits, punctuation symbols and control characters are represented in seven bits by a number from 0 to 127. An eighth bit is often added to allow extra characters.

bibliography. A list of publications on a particular topic, or the reference list of a book.

BibTEX. A program that cooperates with LATEX in the preparation of reference lists. It makes use of `bib` files, which are databases of references in BibTEX format.

citation. A reference in the text to a publication or other source, usually one that is listed in the references.

compositor. The person who typesets the text (especially in traditional printing).

Computing Reviews Classification System. A classification system for computer science. An example of an entry is G.1.3 [Numerical Analysis]: Numerical Linear Algebra—sparse and very large systems.

conjecture. A statement that the author believes to be true but for which a proof or disproof has not been found.

copy editor. A person who prepares a manuscript for typesetting by checking and correcting grammar, punctuation, spelling, style, consistency and other details.

corollary. A direct or easy consequence of a lemma, theorem or proposition.

Current Contents. A publication from the Institute for Scientific Information that provides a weekly list of journal contents pages. The Physical Sciences edition is the one in which mathematics and computer science journals appear.

CTAN. The Comprehensive TEX Archive Network: a network of ftp servers that hold up-to-date copies of all the public domain versions of TEX, LATEX, and related macros and programs.

DOS. Disk Operating System. Usually refers to MS-DOS (Microsoft Disk Operating System), which is a computer operating systems for personal computers (PCs).

Festschrift (or festschrift). (German) A collection of writings published in honour of a scholar.

folio. A printed page number. Also a sheet of a manuscript.

ftp. File transfer protocol. A protocol for file transfer between different computers on the Internet network. Also a program for transferring files using this protocol.

Harvard system. A system of citation by author name and year, e.g., "see Knuth (1986)".

hypertext. On-line text with pointers to other text. For example, a paper provided on a Web page in hypertext format may allow you to link directly to references in the bibliography that are themselves available on the Web.

hypothesis. A statement taken as a basis for further reasoning.

IMA. The Institute of Mathematics and Its Applications, Southend-on-Sea, England. Also the Institute for Mathematics and Its Applications, University of Minnesota, Minneapolis, USA.

Internet. The worldwide network of interconnected computer networks. It provides electronic mail, file transfer, news, remote login, and other services.

ISBN. International Standard Book Number.

ISSN. International Standard Serial Number.

LaTeX. A macro package for TeX that simplifies the production of papers, books and letters, and emphasizes the logical structure of a document. It permits automatic cross-referencing and has commands for drawing pictures.

lemma. An auxiliary result needed in the proof of a theorem or proposition. May also be an independent result that does not merit the title theorem.

LMS. The London Mathematical Society.

MAA. The Mathematical Association of America.

macro. In computing, a shorthand notation for specifying a sequence of operations. For example, in typesetting this book in LaTeX I used the definition `\def\mw{mathematical writing}` and typed `\mw` whenever I wanted the phrase "mathematical writing" to appear.

MakeIndex. A program that makes an index for a LaTeX document.

manuscript. Literally, a handwritten document. More generally, any unpublished document, particularly one submitted for publication.

managing editor. A person who is in charge of the editorial activities of a publication and who supervises a group of editors.

Mathematical Reviews. A monthly review publication run by the American Mathematical Society (AMS) and first published in 1940. Each listed paper is accompanied by a review or a reprint of the paper's abstract.

Mathematics Subject Classifications. A classification scheme published in *Mathematical Reviews* that divides mathematics into 61 sections numbered between 0 and 94, further divided into many subsections. A typical entry is 65F05 (direct methods for solving linear systems).

netlib. An electronic repository of public domain mathematical software for the scientific computing community.

offprint. See reprint.

page charges. Charges levied by a publisher to offset the cost of publishing an article. In mathematics journals payment is usually optional.

PDF. Portable Document Format (PDF), developed by Adobe Systems, Inc. and based on PostScript. Can be read using the Adobe Acrobat software.

peer review. Refereeing done by peers of the author (people working in the same area). Should perhaps be called "peer refereeing", but "peer review" is standard.

poster. A display of graphics and text that summarizes a piece of work. It usually comprises sheets of paper attached to a poster board.

proceedings. A collection of papers describing the work presented at a conference or workshop. Also may be a title for a journal: for example, *The Proceedings of the American Mathematical Society.*

proofreading. The process of checking proofs for errors (usually by comparing them with an original) and marking the errors with standard proofreading symbols.

proofs. Typeset material ready for checking and correction.

proposition. Same meaning as a theorem (but possibly regarded as a lesser result).

PostScript. A page description language developed by Adobe Systems, Inc. Now a standard format in which to provide documents on the World Wide Web.

referee. A person who advises an editor on the suitability of a manuscript for publication.

references. The list of publications cited in the text, or those publications themselves.

reprint. A separate printing of an article that appeared in a book or journal. Often, a limited number are supplied free of charge to the author.

reviewer. A person who reviews previously published or completed work. Sometimes used incorrectly as a synonym for referee.

running head. An abbreviated title that appears in the headline of pages in a published paper.

Science Citation Index. A publication from the Institute for Scientific Information that records all papers that reference an earlier paper, across all science subjects. It covers the period from 1945 to the present.

SIAM. The Society for Industrial and Applied Mathematics.

technical report. A document published by an organization for external circulation, usually as part of a series.

TeX. A system for computer typesetting of mathematics, developed by Donald Knuth at Stanford University. Also used as a verb: "to TeX a paper" is to typeset the paper in TeX.

theorem. A major result of independent interest.

thesaurus. A list of words in which each word is followed by a list of words of similar meaning or sense. The main list may be arranged by meaning (*Roget's Thesaurus*) or alphabetically (most other thesauruses).

title. "The fewest possible words that adequately describe the contents of a paper, book, poster, etc." [68]

Unix. A computer operating system developed at Bell Laboratories. Widely used on workstations and supercomputers.

URL. Uniform resource locator. A URL is the address of an object on the World Wide Web.

widow. A short last line of a paragraph appearing at the top of a page.

World Wide Web. The handbook to the Web browser Netscape explains that "The World Wide Web (WWW or Web) is one facet of the Internet consisting of client and server computers handling multimedia documents."

Bibliography

[1] J. C. Abbott, editor. *The Chauvenet Papers: A Collection of Prize-Winning Expository Papers in Mathematics, Volumes 1 and 2.* Mathematical Association of America, Washington, D.C., 1978.

[2] Milton Abramowitz and Irene A. Stegun, editors. *Handbook of Mathematical Functions with Formulas, Graphs and Mathematical Tables*, volume 55 of *Applied Mathematics Series*. National Bureau of Standards, Washington, D.C., 1964. Reprinted by Dover, New York.

[3] Forman S. Acton. *Numerical Methods That Work.* Harper and Row, New York, 1970. xviii+541 pp. Reprinted by Mathematical Association of America, Washington, D.C., with new preface and additional problems, 1990. ISBN 0-88385-450-3.

[4] Alfred V. Aho, Brian W. Kernighan, and Peter J. Weinberger. *The AWK Programming Language*. Addison-Wesley, Reading, MA, USA, 1988. x+210 pp. ISBN 0-201-07981-X.

[5] D. J. Albers and G. L. Alexanderson, editors. *Mathematical People: Profiles and Interviews*. Birkhäuser, Boston, 1985.

[6] *The American Heritage College Dictionary.* Third edition, Houghton Mifflin, Boston, 1997. xxxiv+1630 pp. ISBN 0-395-67161-2.

[7] *The American Heritage Dictionary of the English Language.* Third edition, Houghton Mifflin, Boston, 1996. xliv+2140 pp. ISBN 0-395-44895-6.

[8] American Mathematical Society. $\mathcal{AMS}$-$\LaTeX$ *Version 1.1 User's Guide.* Providence, RI, USA, 1991.

[9] Robert R. H. Anholt. *Dazzle 'em with Style: The Art of Oral Scientific Presentation.* W. H. Freeman, New York, 1994. xiii+200 pp. ISBN 0-7167-2583-5.

[10] Anonymous. Next slide please. *Nature*, 272(5656):743, 1978.

[11] Don Aslett. *Is There a Speech Inside You?* Writer's Digest Books, Cincinnati, Ohio, 1989. 135 pp. ISBN 0-89879-361-0.

[12] David H. Bailey. Twelve ways to fool the masses when giving performance results on parallel computers. *Supercomputer Rev.*, August:54–55, 1991. Also in *Supercomputer*, Sept. 1991, pp. 4–7.

[13] Sheridan Baker. *The Practical Stylist.* Sixth edition, Harper and Row, New York, 1985. xii+290 pp. ISBN 0-06-040439-6.

269

[14] Robert Barrass. *Scientists Must Write: A Guide to Better Writing for Scientists, Engineers and Students.* Chapman and Hall, London, 1978. xiv+176 pp. ISBN 0-412-15430-7.

[15] Robert Barrass. *Students Must Write: A Guide to Better Writing in Course Work and Examinations.* Methuen, London, 1982. ix+149 pp. ISBN 0-416-33620-5.

[16] Floyd K. Baskette, Jack Z. Sissors, and Brian S. Brooks. *The Art of Editing.* Fifth edition, Macmillan, New York, 1992. viii+518 pp. ISBN 0-02-306295-9.

[17] Nelson H. F. Beebe. Bibliography prettyprinting and syntax checking. *TUGboat*, 14(4):395–419, 1993.

[18] David F. Beer, editor. *Writing and Speaking in the Technology Professions: A Practical Guide.* IEEE Press, New York, 1992. ISBN 0-87942-284-X.

[19] Albert H. Beiler. *Recreations in the Theory of Numbers: The Queen of Mathematics Entertains.* Dover, New York, 1966. xviii+349 pp. ISBN 0-486-21096-0.

[20] Jon L. Bentley. *Programming Pearls.* Addison-Wesley, Reading, MA, USA, 1986. viii+195 pp. ISBN 0-201-10331-1.

[21] Jon L. Bentley. *More Programming Pearls: Confessions of a Coder.* Addison-Wesley, Reading, MA, USA, 1988. viii+207 pp. ISBN 0-201-11889-0.

[22] Jon L. Bentley and Brian W. Kernighan. Tools for printing indexes. *Electronic Publishing*, 1(1):3–17, 1988. Also Computer Science Technical Report No. 128, AT&T Bell Laboratories, Murray Hill, NJ, October 1986.

[23] Abraham Berman and Robert J. Plemmons. *Nonnegative Matrices in the Mathematical Sciences.* Society for Industrial and Applied Mathematics, Philadelphia, PA, 1994. xx+340 pp. Corrected republication, with supplement, of work first published in 1979 by Academic Press. ISBN 0-89871-321-8.

[24] André Bernard, editor. *Rotten Rejections: A Literary Companion.* Penguin, New York, 1990. 101 pp. ISBN 0-14-014786-1.

[25] Theodore M. Bernstein. *The Careful Writer: A Modern Guide to English Usage.* Atheneum, New York, 1965. xviii+487 pp. ISBN 0-689-70555-7.

[26] Theodore M. Bernstein. *Miss Thistlebottom's Hobgoblins: The Careful Writer's Guide to the Taboos, Bugbears and Outmoded Rules of English Usage.* Farrar, Straus and Giroux, New York, 1971. 260 pp. Reprinted by Simon and Schuster, New York, 1984. ISBN 0-671-50404-5.

[27] Theodore M. Bernstein. *Dos, Don'ts and Maybes of English Usage.* Barnes and Noble, New York, 1977. 250 pp. ISBN 0-88029-944-4.

[28] Theodore M. Bernstein. Punctuation. *IEEE Trans. Prof. Commun.*, PC-20 (1):38–44, 1977. Reprinted from [25].

[29] Cicely Berry. *Your Voice and How to Use It Successfully.* Harrap, London, 1975. 160 pp. ISBN 0-245-52886-5.

[30] Ambrose Bierce. *Write it Right: A Little Blacklist of Literary Faults.* The Union Library Association, New York, 1937. Reprinted in [26, pp. 209–253].

[31] Lennart Björck, Michael Knight, and Eleanor Wikborg. *The Writing Process: Composition Writing for University Students.* Second edition, Studentlitteratur, Lund, Sweden and Chartwell Bratt, Bromley, Kent, England, 1990. ISBN 91 44 28222 2 and 0 86238 300 5.

[32] *Bloomsbury Thesaurus.* Bloomsbury, London, 1993. xxx+1569 pp. ISBN 0-7475-1226-4.

[33] Ralph P. Boas. Can we make mathematics intelligible? *Amer. Math. Monthly,* 88:727–731, 1981.

[34] Béla Bollobás, editor. *Littlewood's Miscellany.* Cambridge University Press, 1986. 200 pp. First published in 1953 by Methuen as *A Mathematician's Miscellany.* ISBN 0-521-33702-X.

[35] Larry S. Bonura. *The Art of Indexing.* Wiley, New York, 1994. xxii+233 pp. ISBN 0-471-01449-4.

[36] Vernon Booth. *Communicating in Science: Writing a Scientific Paper and Speaking at Scientific Meetings.* Second edition, Cambridge University Press, 1993. xvi+78 pp. ISBN 0-521-42915-3.

[37] L. A. Brankin and A. M. Mumford. File formats for computer graphics: Unraveling the confusion. Technical Report TR1/92, NAG Ltd., Oxford, June 1992.

[38] Mary Helen Briscoe. *Preparing Scientific Illustrations: A Guide to Better Posters, Presentations, and Publications.* Second edition, Springer-Verlag, New York, 1996. xii+204 pp. ISBN 0-387-94581-4.

[39] William Broad and Nicholas Wade. *Betrayers of the Truth: Fraud and Deceit in Science.* Oxford University Press, 1982. 256 pp. ISBN 0-19-281889-9.

[40] Shirley V. Browne, Jack J. Dongarra, Stan C. Green, Keith Moore, Thomas H. Rowan, and Reed C. Wade. Netlib services and resources. Report ORNL/TM-12680, Oak Ridge National Laboratory, Oak Ridge, TN, USA, April 1994. 42 pp.

[41] Bill Bryson. *The Penguin Dictionary of Troublesome Words.* Second edition, Penguin, London, 1987. 192 pp. ISBN 0-14-051200-4.

[42] Bill Bryson. *Mother Tongue: English and How It Got That Way.* Avon Books, New York, 1990. 270 pp. ISBN 0-380-71543-0.

[43] Robert Burchfield. *Unlocking the English Language.* Faber and Faber, London, 1989. xv+202 pp. ISBN 0-571-14416-0.

[44] David M. Burton. *Elementary Number Theory.* Allyn and Bacon, Boston, 1980. ix+390 pp. ISBN 0-205-06978-9.

[45] Judith Butcher. *Copy-Editing: The Cambridge Handbook for Editors, Authors and Publishers*. Third edition, Cambridge University Press, 1992. xii+471 pp. ISBN 0-521-40074-0.

[46] Tony Buzan. *Use Your Head*. BBC Publications, London, 1982. 156 pp. ISBN 0-563-16552-9.

[47] Bill Buzbee. Poisson's equation revisited. *Current Contents*, 36:8, 1992.

[48] George D. Byrne. How to improve technical presentations. *SIAM News*, 20:10–11, January 1987.

[49] Florian Cajori. *A History of Mathematical Notations. Two Volumes Bound as One. Volume I: Notations in Elementary Mathematics. Volume II: Notations Mainly in Higher Mathematics*. Dover, New York, 1993. xxviii+820 pp. Reprint of works originally published in 1928 and 1929 by The Open Court Publishing Company, Chicago. ISBN 0-486-67766-4.

[50] James Calnan and Andras Barabas. *Speaking at Medical Meetings: A Practical Guide*. Second edition, William Heinemann Medical, London, 1981. xii+184 pp. ISBN 0-433-05001-2.

[51] Debra Cameron and Bill Rosenblatt. *Learning GNU Emacs*. O'Reilly & Associates, Sebastopol, CA, 1991. xxvii+411 pp. ISBN 0-937175-84-6.

[52] G. V. Carey. *Mind the Stop: A Brief Guide to Punctuation with a Note on Proof-Correction*. Second edition, Penguin, London, 1958. 126 pp. Reprinted 1976. ISBN 0-14-051072-9.

[53] David P. Carlisle and Nicholas J. Higham. LaTeX 2_ε: Should you upgrade to it? *SIAM News*, 29(1):12, 1996.

[54] *The Chambers Dictionary*. Chambers, Edinburgh, 1993. xviii+2062 pp. ISBN 0-550-10255-8.

[55] T. W. Chaundy, P. R. Barrett, and Charles Batey. *The Printing of Mathematics: Aids for Authors and Editors and Rules for Compositors and Readers at the University Press, Oxford*. Oxford University Press, 1954.

[56] Pehong Chen and Michael A. Harrison. Index preparation and processing. *Software—Practice and Experience*, 18(9):897–915, 1988.

[57] Lorinda L. Cherry. Writing tools. *IEEE Trans. Communications*, COM-30 (1):100–105, 1982.

[58] *The Chicago Manual of Style*. Fourteenth edition, University of Chicago Press, Chicago and London, 1993. ix+921 pp. ISBN 0-226-10389-7.

[59] *Collins Cobuild English Dictionary*. Collins, London, 1995. xxxix+1951 pp. ISBN 0-00-370941-8.

[60] *Collins English Dictionary*. Third edition, HarperCollins, Glasgow, 1991. xxxi+1791 pp. ISBN 0-00-433286-5.

[61] *Collins Plain English Dictionary*. HarperCollins, London, 1996. 758 pp. ISBN 0-00-375056-6.

[62] Bruce M. Cooper. *Writing Technical Reports*. Penguin, London, 1964. 188 pp. Reprinted 1986. ISBN 0-14-020676-0.

[63] Michael Crichton. Medical obfuscation: Structure and function. *The New England Journal of Medicine*, 293(24):1257–1259, 1975.

[64] Francis Crick. The double helix: A personal view. *Nature*, 248:766–769, 1974.

[65] H. Crowder, R. S. Dembo, and J. M. Mulvey. On reporting computational experiments with mathematical software. *ACM Trans. Math. Software*, 5: 193–203, 1979.

[66] David Crystal. *The English Language*. Penguin, London, 1988. 124 pp. ISBN 0-14-022730-X.

[67] P. J. Davis. Fidelity in mathematical discourse: Is one and one really two? *Amer. Math. Monthly*, 79:252–263, 1972.

[68] Robert A. Day. *How To Write and Publish a Scientific Paper*. Fourth edition, Cambridge University Press, and Oryx Press, Phoenix, Arizona, 1994. xiv+223 pp. ISBN 0-521-55898-0.

[69] Robert A. Day. *Scientific English: A Guide for Scientists and Other Professionals*. Second edition, Oryx Press, Phoenix, Arizona, 1995. xii+148 pp. ISBN 0-89774-989-8.

[70] J. T. Dillon. The emergence of the colon: An empirical correlate of scholarship. *American Psychologist*, 36(8):879–884, 1981.

[71] Bernard Dixon. Sciwrite. *Chemistry in Britain*, 9(1):70–72, 1973.

[72] Jack J. Dongarra and Eric Grosse. Distribution of mathematical software via electronic mail. *Comm. ACM*, 30(5):403–407, 1987.

[73] Jack J. Dongarra and Bill Rosener. NA-NET: Numerical analysis NET. Technical Report CS-91-146, Department of Computer Science, University of Tennessee, Knoxville, September 1991. 21 pp.

[74] Susan Dressel and Joe Chew. Authenticity beats eloquence. *IEEE Trans. Prof. Commun.*, PC-30:82–83, 1987.

[75] Freeman Dyson. George Green and physics. *Physics World*, 6(8):33–38, 1993.

[76] Hans F. Ebel, Claus Bliefert, and William E. Russey. *The Art of Scientific Writing: From Student Reports to Professional Publications in Chemistry and Related Fields*. VCH Publishers, New York, 1987. ISBN 0-89573-645-4.

[77] Anne Eisenberg. *Guide to Technical Editing: Discussion, Dictionary, and Exercises*. Oxford University Press, New York, 1992. ix+182 pp. ISBN 0-19-506306-6.

[78] J. R. Ewer and G. Latorre. *A Course in Basic Scientific English*. Longman, Harlow, Essex, 1969. ISBN 0-582-52009-6.

[79] John Ewing, editor. *A Century of Mathematics Through the Eyes of the Monthly.* Mathematical Association of America, Washington, D.C., 1994. xi+323 pp. ISBN 0-88385-459-7.

[80] Harley Flanders. Manual for Monthly authors. *Amer. Math. Monthly,* 78: 1–10, 1971.

[81] Rudolf Flesch. *The Art of Plain Talk.* Harper and Brothers, New York, 1946. xiii+210 pp.

[82] G. E. Forsythe. Suggestions to students on talking about mathematics papers. *Amer. Math. Monthly,* 64:16–18, 1957. Reprinted in [79].

[83] H. W. Fowler. *A Dictionary of Modern English Usage.* Second edition, Oxford University Press, 1968. Revised by Sir Ernest Gowers.

[84] H. W. Fowler and F. G. Fowler. *The King's English.* Third edition, Oxford University Press, 1931. 382 pp. Reprinted 1990. ISBN 0-19-881330-9.

[85] James Franklin and Albert Daoud. *Introduction to Proofs in Mathematics.* Prentice-Hall, Englewood Cliffs, NJ, USA, 1988. vii+175 pp. ISBN 0-13-474313-X.

[86] Daniel H. Freeman, Jr., María Elena González, David C. Hoaglin, and Beth A. Kilss. Presenting statistical papers. *The American Statistician,* 37 (2):106–110, 1983.

[87] Matthew P. Gaffney and Lynn Arthur Steen. *Annotated Bibliography of Expository Writing in the Mathematical Sciences.* Mathematical Association of America, Washington, D.C., 1976. xi+282 pp. ISBN 0-88385-422-8.

[88] Eugene Garfield. "Science Citation Index"—A new dimension in indexing. *Science,* 144(3619):649–654, 1964.

[89] Eugene Garfield. Citation analysis as a tool in journal evaluation. *Science,* 178:471–479, 1972.

[90] Eugene Garfield. Significant journals of science. *Nature,* 264:609–615, 1976.

[91] Eugene Garfield. *Citation Indexing: Its Theory and Application in Science, Technology, and Humanities.* John Wiley, New York, 1979. xxi+274 pp. ISBN 0-471-02559-3.

[92] Eugene Garfield. Journal citation studies. 36. Pure and applied mathematics journals: What they cite and vice versa. *Current Contents,* 5(15):5–10, 1982.

[93] Eugene Garfield. The 100 most-cited papers ever and how we select citation classics. *Current Contents,* 23, 1984.

[94] Eugene Garfield. *The Awards of Science and Other Essays.* Essays of an Information Scientist: 1984. ISI Press, Philadelphia, 1985.

[95] Eugene Garfield. Journal editors awaken to the impact of citation errors. How we control them at ISI. *Current Contents,* 41:5–13, 1990.

[96] Eugene Garfield. The most-cited papers of all time, SCI 1945–1988. Part 1A. The SCI top 100—will the Lowry method ever be obliterated? *Current Contents*, 7:3–14, 1990.

[97] Eugene Garfield. The most-cited papers of all time, SCI 1945–1988. Part 1B. Superstars new to the SCI top 100. *Current Contents*, 8:3–13, 1990.

[98] Eugene Garfield. The most-cited papers of all time, SCI 1945–1988. Part 3. Another 100 from the Citation Classics Hall of Fame. *Current Contents*, 34:3–13, 1990.

[99] Eugene Garfield. How to use the science citation index (SCI). In *SCI Science Citation Index 1991. Guide and List of Source Publications*, ISI Press, Philadelphia, PA, 1991, pages 26–33. Reprinted from *Current Contents*, 9: 5–14, 1983.

[100] Eugene Garfield. The impact factor. *Current Contents*, 34(25):3–7, 1994.

[101] Rowan Garnier and John Taylor. *100% Mathematical Proof.* Wiley, Chichester, UK, 1996. viii+317 pp. ISBN 0-471-96199-X.

[102] Robert V. Garver. Presenting the peer paper. *IEEE Trans. Prof. Commun.*, PC-23(1):18–22, 1980.

[103] C. William Gear. Inside SIAM's mysterious journal publication process. *SIAM News*, 24(2):6, March 1991.

[104] Leonard Gillman. *Writing Mathematics Well: A Manual for Authors.* The Mathematical Association of America, Washington, D.C., 1987. ix+49 pp. ISBN 0-88385-443-0.

[105] Leonard Gillman. Paul Halmos's expository writing. In *Paul Halmos: Celebrating 50 Years of Mathematics*, John H. Ewing and F. W. Gehring, editors, Springer-Verlag, Berlin, 1991, pages 33–48.

[106] Leon J. Gleser. Some notes on refereeing. *The American Statistician*, 40 (4):310–312, 1986.

[107] *GNU Emacs Manual, Emacs Version 19.* Free Software Foundation, 59 Temple Place, Suite 330, Boston, MA 02111-1307, USA. Available on-line with the GNU Emacs distribution.

[108] Gene H. Golub and Charles F. Van Loan. *Matrix Computations.* Third edition, Johns Hopkins University Press, Baltimore, MD, USA, 1996. xxvii+694 pp. ISBN 0-8018-5413-X (hardback), 0-8018-5414-8 (paperback).

[109] Michael Goossens, Frank Mittelbach, and Alexander Samarin. *The LaTeX Companion.* Addison-Wesley, Reading, MA, USA, 1993. xxx+528 pp. ISBN 0-201-54199-8.

[110] Michael Goossens, Sebastian Rahtz, and Frank Mittelbach. *The LaTeX Graphics Companion: Illustrating Documents with TeX and PostScript.* Addison-Wesley, Reading, MA, USA, 1997. xxv+554 pp. ISBN 0-201-85469-4.

[111] Karen Elizabeth Gordon. *The Transitive Vampire: A Handbook of Grammar for the Innocent, the Eager, and the Doomed.* Revised and expanded edition, Times Books, New York, 1984. x+149 pp. ISBN 0-8129-1101-6.

[112] Karen Elizabeth Gordon. *The New Well-Tempered Sentence: A Punctuation Handbook for the Innocent, the Eager, and the Doomed.* Revised and expanded edition, Ticknor and Fields, New York, 1993. x+148 pp. ISBN 0-395-62883-0.

[113] Calvin R. Gould. The overhead projector. *IEEE Trans. Prof. Commun.*, PC-15(1):2–6, 1972.

[114] Calvin R. Gould. Visual aids—how to make them positively legible. *IEEE Trans. Prof. Commun.*, PC-16(2):35–38, 1973.

[115] Sir Ernest Gowers. *The Complete Plain Words.* Third edition, Penguin, London, 1986. vi+288 pp. Revised by Sidney Greenbaum and Janet Whitcut. ISBN 0-14-051199-7.

[116] Ronald L. Graham, Donald E. Knuth, and Oren Patashnik. *Concrete Mathematics: A Foundation for Computer Science.* Second edition, Addison-Wesley, Reading, MA, USA, 1994. xiii+657 pp. ISBN 0-201-55802-5.

[117] Martin W. Gregory. The infectiousness of pompous prose. *Nature*, 360: 11–12, 1992.

[118] David F. Griffiths and Desmond J. Higham. *Learning LATEX.* Society for Industrial and Applied Mathematics, Philadelphia, PA, 1997. 84 pp. ISBN 0-89871-383-8.

[119] R. Grone, C. R. Johnson, E. M. Sá, and H. Wolkowicz. Normal matrices. *Linear Algebra and Appl.*, 87:213–225, 1987.

[120] Jerrold W. Grossman. Paul Erdös: The master of collaboration. In *The Mathematics of Paul Erdös II*, Ronald L. Graham and Jaroslav Nešetřil, editors, Springer-Verlag, Berlin, 1997, pages 467–475.

[121] Paul R. Halmos. How to write mathematics. *Enseign. Math.*, 16:123–152, 1970. Reprinted in [257] and [245].

[122] Paul R. Halmos. *Finite-Dimensional Vector Spaces.* Springer-Verlag, New York, 1974. viii+200 pp. ISBN 0-387-90093-4.

[123] Paul R. Halmos. How to talk mathematics. *Notices Amer. Math. Soc.*, 21 (3):155–158, 1974. Reprinted in [245].

[124] Paul R. Halmos. What to publish. *Amer. Math. Monthly*, 82(1):14–17, 1975. Reprinted in [245].

[125] Paul R. Halmos. *A Hilbert Space Problem Book.* Second edition, Springer-Verlag, Berlin, 1982.

[126] Paul R. Halmos. Think it gooder. *The Mathematical Intelligencer*, 4(1): 20–21, 1982.

[127] Paul R. Halmos. *I Want to Be a Mathematician: An Automathography in Three Parts.* Springer-Verlag, New York, 1985. xv+421 pp. ISBN 0-88385-445-7.

[128] Paul R. Halmos. Some books of auld lang syne. In *A Century of Mathematics in America, Part I*, Peter Duren, Richard A. Askey, and Uta C. Merzbach, editors, American Mathematical Society, Providence, RI, USA, 1988, pages 131–174.

[129] Sven Hammarling and Nicholas J. Higham. How to prepare a poster. *SIAM News*, 29(4):20, 19, May 1996.

[130] Leonard Montague Harrod, editor. *Indexers on Indexing: A Selection of Articles Published in* The Indexer. R. K. Bowker, London, 1978. x+430 pp. ISBN 0-8352-1099-5.

[131] Horace Hart. *Hart's Rules for Compositors and Readers at the University Press Oxford.* Thirty-ninth edition, Oxford University Press, 1983. xi+182 pp. First edition 1893. ISBN 0-19-212983-X.

[132] James Hartley. Eighty ways of improving instructional text. *IEEE Trans. Prof. Commun.*, PC-24:17–27, 1981.

[133] James Hartley. The role of colleagues and text-editing programs in improving text. *IEEE Trans. Prof. Commun.*, PC-27:42–44, 1984.

[134] James Hartley. *Designing Instructional Text.* Second edition, Kogan Page, London, 1985. 175 pp. ISBN 0-85038-943-7.

[135] Edward F. Hartree. Ethics for authors: A case history of acrosin. *Perspectives in Biology and Medicine*, 20:82–91, 1976.

[136] J. B. Heaton and N. D. Turton. *Longman Dictionary of Common Errors.* Longman, Harlow, Essex, 1987. ISBN 0-582-96410-5.

[137] John L. Hennessy and David A. Patterson. *Computer Architecture: A Quantitative Approach.* Morgan Kaufmann, San Mateo, CA, USA, 1990. xxviii+594+appendices pp. ISBN 1-55860-188-0.

[138] A. J. Herbert. *The Structure of Technical English.* Longman, Harlow, Essex, 1965. ISBN 0-582-52523-3.

[139] Desmond J. Higham. More commandments of good writing. Manuscript, Department of Mathematics and Computer Science, University of Dundee, UK, November 1992.

[140] Nicholas J. Higham. Algorithm 694: A collection of test matrices in MATLAB. *ACM Trans. Math. Software*, 17(3):289–305, September 1991.

[141] Nicholas J. Higham. Which dictionary for the mathematical scientist? *IMA Bulletin*, 30(5/6):81–88, 1994.

[142] Philip J. Hills, editor. *Publish or Perish.* Peter Francis Publishers, Berrycroft, Cambridgeshire, UK, 1987. 186 pp. ISBN 1-8701-6700-7.

[143] Alston S. Householder. *The Theory of Matrices in Numerical Analysis.* Blaisdell, New York, 1964. xi+257 pp. Reprinted by Dover, New York, 1975. ISBN 0-486-61781-5.

[144] Kenneth E. Iverson. *A Programming Language.* Wiley, New York, 1962.

[145] Donald D. Jackson. A brief history of scholarly publishing. In [292, pp. 133–134].

[146] R. H. F. Jackson, P. T. Boggs, S. G. Nash, and S. Powell. Guidelines for reporting results of computational experiments. Report of the ad hoc committee. *Math. Prog.*, 49:413–425, 1991.

[147] J. L. Kelley. Writing mathematics. In *Paul Halmos: Celebrating 50 Years of Mathematics*, John H. Ewing and F. W. Gehring, editors, Springer-Verlag, Berlin, 1991, pages 91–96.

[148] Kevin Kelly, editor. *SIGNAL: Communication Tools for the Information Age*. Harmony Books, Crown Publishers, New York. ISBN 0-517-57084-X.

[149] Peter Kenny. *A Handbook of Public Speaking for Scientists and Engineers*. Adam Hilger, Bristol, 1982. xi+181 pp. ISBN 0-85274-553-2.

[150] G. A. Kerkut. Choosing a title for a paper. *Comp. Biochem. Physiol.*, 47A(1):1, 1983.

[151] Brian W. Kernighan and Lorinda L. Cherry. A system for typesetting mathematics. *Comm. ACM*, 18(3):151–157, 1975.

[152] Lester S. King. Medical writing number 7: The opening sentence. *J. Amer. Medical Assoc.*, 202(6):535–536, 1967.

[153] John Kirkman. *Good Style: Writing for Science and Technology*. E & FN Spon (Chapman and Hall), London, 1992. viii+221 pp. ISBN 0-419-17190-8.

[154] Charles Kittel. *Introduction to Solid State Physics*. Fourth edition, Wiley, New York, 1971. xv+766 pp. ISBN 0-471-49021-0.

[155] George R. Klare. *The Measurement of Readability*. Iowa State University Press, Ames, IA, USA, 1963.

[156] G. Norman Knight. Book indexing in Great Britain: A brief history. *The Indexer*, 6(1):14–18, 1968. Reprinted in [130, pp. 9–13].

[157] Donald E. Knuth. *The Art of Computer Programming*. Addison-Wesley, Reading, MA, USA, 1973–1981. Three volumes.

[158] Donald E. Knuth. Mathematical typography. *Bulletin Amer. Math. Soc. (New Series)*, 1(2):337–372, 1979.

[159] Donald E. Knuth. *The Art of Computer Programming, Volume 2, Seminumerical Algorithms*. Second edition, Addison-Wesley, Reading, MA, USA, 1981. xiii+688 pp. ISBN 0-201-03822-6.

[160] Donald E. Knuth. *The METAFONT Book*. Addison-Wesley, Reading, MA, USA, 1986. xi+361 pp. ISBN 0-201-13444-6.

[161] Donald E. Knuth. *The TEXbook*. Addison-Wesley, Reading, MA, USA, 1986. ix+483 pp. ISBN 0-201-13448-9.

[162] Donald E. Knuth. *3:16 Bible Texts Illuminated*. A-R Editions, Madison, WI, 1991. 268 pp. ISBN 0-89579-252-4.

[163] Donald E. Knuth. Two notes on notation. *Amer. Math. Monthly*, 99(5): 403–422, 1992.

[164] Donald E. Knuth, Tracy Larrabee, and Paul M. Roberts. *Mathematical Writing*. MAA Notes Number 14. Mathematical Association of America, Washington, D.C., 1989. 115 pp. Also Report STAN-CS-88-1193, Department of Computer Science, Stanford University, Stanford, CA, USA, January 1988. ISBN 0-88385-063-X.

[165] Kodak Limited. Let's stamp out awful lecture slides. Kodak publication S-22(H), April 1979.

[166] Helmut Kopka and Patrick W. Daly. *A Guide to LaTeX 2ε: Document Preparation for Beginners and Advanced Users*. Second edition, Addison-Wesley, Wokingham, England, 1995. x+554 pp. ISBN 0-201-42777-X.

[167] Steven G. Krantz. *A Primer of Mathematical Writing: Being a Disquisition on Having Your Ideas Recorded, Typeset, Published, Read, and Appreciated*. American Mathematical Society, Providence, RI, USA, 1997. xv+223 pp. ISBN 0-8218-0635-1.

[168] Ed Krol. *The Whole Internet User's Guide & Catalog*. Second edition, O'Reilly & Associates, Sebastopol, CA, USA, 1994. xxv+543 pp. ISBN 1-56592-063-5.

[169] Marcel C. LaFollette. *Stealing into Print: Fraud, Plagiarism, and Misconduct in Scientific Publishing*. University of California Press, Berkeley, CA, 1992. viii+293 pp. ISBN 0-520-07831-4.

[170] David Lambuth et al. *The Golden Book on Writing*. Penguin, 1964. xiv+81 pp. ISBN 0-14-046263-5.

[171] Leslie Lamport. Document production: Visual or logical? *Notices Amer. Math. Soc.*, 34:621–624, 1987.

[172] Leslie Lamport. *LaTeX: A Document Preparation System. User's Guide and Reference Manual*. Second edition, Addison-Wesley, Reading, MA, USA, 1994. xvi+272 pp. ISBN 0-201-52983-1.

[173] Leslie Lamport. How to write a proof. *Amer. Math. Monthly*, 102(7): 600–608, 1995.

[174] Kenneth K. Landes. A scrutiny of the abstract. II. *Bull. Amer. Assoc. Petroleum Geologists*, 50(9):1992, 1966.

[175] Richard A. Lanham. *Revising Prose*. Third edition, Macmillan, New York, 1992. xi+123 pp. ISBN 0-02-367445-8.

[176] Tracey LaQuey and Jeanne C. Ryer. *The Internet Companion: A Beginner's Guide to Global Networking*. Addison-Wesley, Reading, MA, USA, 1993. x+196 pp. ISBN 0-201-62224-6.

[177] Uri Leron. Structuring mathematical proofs. *Amer. Math. Monthly*, 90(3): 174–185, 1983.

[178] Xia Li and Nancy B. Crane. *Electronic Styles: A Handbook for Citing Electronic Information.* Information Today, Inc., Medford, NJ, USA, 1996. xviii+213 pp. ISBN 1-57387-027-7.

[179] Dennis V. Lindley. Refereeing. *The Mathematical Intelligencer*, 6(2):56–60, 1984.

[180] Stephen Lock, editor. *How to Do It.* Second edition, British Medical Association, London, 1985. ISBN 0-7279-0186-9.

[181] *Longman Dictionary of Contemporary English.* Third edition, Longman, Harlow, Essex, 1995. xxii+1668 pp. ISBN 0-582-23750-5.

[182] *Longman Dictionary of the English Language.* New edition, Longman, Harlow, Essex, 1991. xxv+1890 pp. ISBN 0-582-07038-4.

[183] Harry Lorayne. *How to Develop a Super-Power Memory.* Signet, New York, 1974. xii+180 pp. ISBN 0-451-12941-5.

[184] Harry Lorayne and Jerry Lucas. *The Memory Book.* Wyndham, London, 1974. 207 pp. ISBN 0-352-39856-6.

[185] Beth Luey. *Handbook for Academic Authors.* Third edition, Cambridge University Press, 1995. ISBN 0-521-49892-9.

[186] Nina H. Macdonald, Lawrence T. Frase, Patricia S. Gingrich, and Stacey A. Keenan. The Writer's Workbench: Computer aids for text analysis. *IEEE Trans. Communications*, COM-30(1):105–110, 1982.

[187] A. J. MacGregor. Graphics simplified: Charts and graphs. *Scholarly Publishing*, 8(2):151–164, 1977.

[188] A. J. MacGregor. Graphics simplified: Preparing charts and graphs. *Scholarly Publishing*, 8(3):257–274, 1977.

[189] N. J. Mackintosh, editor. *Cyril Burt: Fraud or Framed?* Oxford University Press, 1995. vii+156 pp. ISBN 0-19-852336-X.

[190] Donald S. MacQueen. *Using Numbers in English: A Reference Guide for Swedish Speakers Including Basic Terminology for Describing Graphs.* Studentlitteratur, Lund, Sweden and Chartwell Bratt, Bromley, Kent, England, 1990. ISBN 91-4431921-5 and 0-862382645.

[191] John Maddox. Must science be impenetrable? *Nature*, 305(6):477–478, 1983.

[192] Thomas Mallon. *Stolen Words: Forays Into the Origins and Ravages of Plagiarism.* Penguin, London, 1989. xiv+300 pp. ISBN 0-14-014440-4.

[193] Frank T. Manheim. The scientific referee. *IEEE Trans. Prof. Commun.*, PC-18(3):190–195, 1975.

[194] S. D. Mason. Oral examination procedure. In [235, pp. 160–161].

[195] Diane L. Matthews. The scientific poster: Guidelines for effective visual communication. *Technical Communication*, 37(3):225–232, 1990.

[196] Thomas H. Maugh, II. Poster sessions: A new look at scientific meetings. *Science*, 184:1361, June 1974.

[197] Stephen B. Maurer. Advice for undergraduates on special aspects of writing mathematics. *PRIMUS (Problems Resources and Issues in Mathematics Undergraduate Studies)*, 1(1):9–28, March 1991.

[198] Glenda M. McClure. Readability formulas: Useful or useless? *IEEE Trans. Prof. Commun.*, PC-30:12–15, 1987.

[199] M. Douglas McIlroy. Development of a spelling list. *IEEE Trans. Communications*, COM-30:91–99, 1982.

[200] N. David Mermin. What's wrong with these equations? *Physics Today*, April:9, 1988. Reprinted in [202].

[201] N. David Mermin. What's wrong with this Lagrangean? *Physics Today*, April:9, 1988. Reprinted with postscript in [202].

[202] N. David Mermin. *Boojums All the Way Through: Communicating Science in a Prosaic Age*. Cambridge University Press, 1990. xxi+309 pp. ISBN 0-521-38880-5.

[203] *Merriam-Webster's Collegiate Dictionary*. Tenth edition, Merriam-Webster, Springfield, MA, USA, 1993. 1559 pp. ISBN 0-87779-708-0.

[204] James A. Michener. *James A. Michener's Writer's Handbook: Explorations in Writing and Publishing*. Random House, New York, 1992. ix+182 pp. ISBN 0-679-74126-7.

[205] Joan P. Mitchell. *The New Writer: Techniques for Writing Well with a Computer*. Microsoft Press, Redmond, WA, USA, 1987. viii+245 pp. ISBN 1-55615-029-6.

[206] R. D. Nelson, editor. *The Penguin Dictionary of Mathematics*. Second edition. Penguin, London, 1998. 350 pp. ISBN 0-14-051342-6.

[207] Maeve O'Connor. *Editing Scientific Books and Journals*. Pitman Medical, Tunbridge Wells, Kent, UK, 1978. ISBN 0-27279517-8.

[208] Maeve O'Connor. *How to Copyedit Scientific Books and Journals*. ISI Press, Philadelphia, PA, 1986. ix+150 pp. ISBN 0-89495-064-9.

[209] Maeve O'Connor. *Writing Successfully in Science*. Chapman and Hall, London, 1991. xi+229 pp. ISBN 0-412-446308.

[210] Maeve O'Connor and F. Peter Woodford. *Writing Scientific Papers in English*. Pitman Medical, Tunbridge Wells, Kent, 1977. vii+108 pp. ISBN 0-272-79515-1.

[211] D. P. O'Leary, G. W. Stewart, and J. S. Vandergraft. Estimating the largest eigenvalue of a positive definite matrix. *Math. Comp.*, 33:1289–1292, 1979.

[212] *Oxford Advanced Learner's Dictionary of Current English*. Fifth edition, Oxford University Press, 1995. x+1428 pp. ISBN 0-19-431422-7.

[213] *The Concise Oxford Dictionary of Current English*. Ninth edition, Oxford University Press, 1995. xxi+1673 pp. ISBN 0-19-861319-9.

[214] *The New Shorter Oxford English Dictionary*. Oxford University Press, 1993. xxvii+3801 pp. ISBN 0-19-861134-X.

[215] *The Oxford English Dictionary*. Second edition, Oxford University Press, 1989. ISBN 0-19-861186-2.

[216] Ian Parberry. A guide for new referees in theoretical computer science. ftp://ftp.unt.edu/ian/guides/referee/manuscript.ps, 1994.

[217] Beresford N. Parlett. *The Symmetric Eigenvalue Problem*. Society for Industrial and Applied Mathematics, Philadelphia, PA, 1998. xxiv+398 pp. Unabridged, amended version of book first published by Prentice-Hall in 1980. ISBN 0-89871-402-8.

[218] Eric Partridge. *Usage and Abusage: A Guide to Good English*. Penguin, London, 1973. 381 pp. ISBN 0-14-051024-9.

[219] Jan A. Pechenik. *A Short Guide to Writing About Biology*. HarperCollins, New York, 1987. xiv+194 pp. ISBN 0-673-39232-5.

[220] John E. Pemberton. *How to Find Out in Mathematics*. Second edition, Pergamon, London, 1969.

[221] Carol Rosenblum Perry. *The Fine Art of Technical Writing*. Blue Heron Publishing, Hillsboro, OR, USA, 1991. 112 pp. ISBN 0-936085-24-X.

[222] H. Pétard. A brief dictionary of phrases used in mathematical writing. *Amer. Math. Monthly*, 73:196–197, 1966. Reprinted in [79].

[223] Ivars Peterson. Searching for new mathematics. *SIAM Review*, 33:37–42, 1991.

[224] James L. Peterson. *Computer Programs for Spelling Correction: An Experiment in Program Design*. Number 96 in Lecture Notes in Computer Science. Springer-Verlag, Berlin, 1980.

[225] James L. Peterson. A note on undetected typing errors. *Comm. ACM*, 29 (7):633–637, 1986.

[226] Estelle M. Phillips and D. S. Pugh. *How to Get a PhD: A Handbook for Students and Their Supervisors*. Second edition, Open University Press, Buckingham, UK, 1994. xiv+203 pp. ISBN 0-335-19214-9.

[227] George Piranian. Say it better. *The Mathematical Intelligencer*, 4(1):17–19, 1982.

[228] George Polya. *How to Solve It: A New Aspect of Mathematical Method*. Second edition, Doubleday, New York, 1957. xxi+253 pp.

[229] Simeon Potter. *Our Language*. Penguin, London, 1976.

[230] W. M. Priestley. Paul Halmos: Words more than numbers. In *Paul Halmos: Celebrating 50 Years of Mathematics*, John H. Ewing and F. W. Gehring, editors, Springer-Verlag, Berlin, 1991, pages 49–69.

[231] D. A. Pyke. Referee a paper. In [180, pp. 215–219].

[232] Randolph Quirk and Gabriele Stein. *English in Use*. Longman, Harlow, Essex, 1990. 262 pp. ISBN 0-582-06613-1.

[233] *Random House Unabridged Dictionary*. Second edition, Random House, New York, 1993. xxxix+2478+32 pp. ISBN 0-679-42917-4.

[234] *Random House Webster's College Dictionary*. Second edition, Random House, New York, 1997. xxxii+1535 pp. ISBN 0-679-45570-1.

[235] *A Random Walk in Science*. An anthology compiled by Robert L. Weber and edited by Eric Mendoza. The Institute of Physics, Bristol and London, 1973. xvii+206 pp. ISBN 0-85498-027-X.

[236] Jerome Irving Rodale. *The Synonym Finder*. Warner Books, New York, 1978. 1361 pp. Completely revised by Laurence Urdang and Nancy LaRoche. ISBN 0-446-37029-0.

[237] Patsy Rodenburg. *The Right to Speak: Working with the Voice*. Methuen, London, 1992. xiv+306 pp. ISBN 0-413-66130-X.

[238] Ervin Y. Rodin. Speed of publication—an editorial. *Computers Math. Applic.*, 24(4):1–2, 1992.

[239] Charles G. Roland. Thoughts about medical writing XXXVII. Verify your references. *Anesthesia and Analgesia ... Current Researches*, 55(5):717–718, 1976.

[240] Richard Rubinstein. *Digital Typography: An Introduction to Type and Composition for Computer System Design*. Addison-Wesley, Reading, MA, USA, 1988. ISBN 0-201-17633-5.

[241] Kjell Erik Rudestam and Rae R. Newton. *Surviving Your Dissertation: A Comprehensive Guide to Content and Process*. Sage Publications, Newbury Park, CA, USA, 91320, 1992. xi+221 pp. ISBN 0-8039-4563-9.

[242] Stephan M. Rudolfer and Peter C. Watson. Table errata. *Math. Comp.*, 59 (206):727, 1992.

[243] William Safire. *Fumblerules: A Lighthearted Guide to Grammar and Good Usage*. Dell Publishing, New York, 1990. 152 pp. ISBN 0-440-21010-0.

[244] David Salomon. *The Advanced TEXbook*. Springer-Verlag, New York, 1995. xx+491 pp. ISBN 0-387-94556-3.

[245] Donald E. Sarason and Leonard Gillman, editors. *P. R. Halmos. Selecta: Expository Writing*. Springer-Verlag, New York, 1983. xix+304 pp. ISBN 0-387-90756-4.

[246] David Louis Schwartz. How to be a published mathematician without trying harder than necessary. In *The Journal of Irreproducible Results: Selected Papers*, George H. Scherr, editor, third edition, 1986, page 205.

[247] Steven Schwartzman. *The Words of Mathematics: An Etymological Dictionary of Mathematical Terms Used in English*. Mathematical Association of America, Washington, D.C., 1994. vii+261 pp. ISBN 0-88385-511-9.

[248] Steven Schwartzman. Number words in English. *College Mathematics Journal*, 26(3):191–195, 1995.

[249] Marjorie E. Skillin, Robert M. Gay, and other authorities. *Words Into Type.* Third edition, Prentice-Hall, Englewood Cliffs, NJ, USA, 1974. xx+583 pp. ISBN 0-13-964262-5.

[250] Alan Jay Smith. The task of the referee. *IEEE Computer*, 23(4):65–71, 1990.

[251] Michael D. Spivak. *The Joy of TEX: A Gourmet Guide to Typesetting with the $\mathcal{A}_{\mathcal{M}}S$-TEX Macro Package.* Second edition, American Mathematical Society, Providence, RI, USA, 1990.

[252] Elsie Myers Stainton. A bag for editors. *Scholarly Publishing*, 8(2):111–119, 1977.

[253] Elsie Myers Stainton. The uses of dictionaries. *Scholarly Publishing*, 11(3): 229–241, 1980.

[254] Elsie Myers Stainton. *The Fine Art of Copyediting.* Columbia University Press, New York, 1991. xi+126 pp. ISBN 0-231-06961-8.

[255] De Witt T. Starnes and Gertrude E. Noyes. *The English Dictionary from Cawdrey to Johnson 1604–1755.* University of North Carolina Press, Chapel Hill, NC, USA, 1946. New edition with an introduction and a select bibliography by Gabriele Stein, John Benjamins Publishing Company, Amsterdam and Philadelphia, 1991. ISBN 90-272-4544-4.

[256] Norman E. Steenrod. How to write mathematics. In [257, pp. 1–17].

[257] Norman E. Steenrod, Paul R. Halmos, Menahem M. Schiffer, and Jean A. Dieudonné. *How to Write Mathematics.* American Mathematical Society, Providence, RI, USA, 1973.

[258] David Sternberg. *How to Complete and Survive a Doctoral Dissertation.* St. Martin's Press, New York, 1981. 231 pp. ISBN 0-312-39606-6.

[259] Andrew Sterrett, editor. *Using Writing to Teach Mathematics.* MAA Notes Number 16. The Mathematical Association of America, Washington, D.C., 1990. xvii+139 pp. ISBN 0-88385-066-4.

[260] Hans J. Stetter. *Analysis of Discretization Methods for Ordinary Differential Equations.* Springer-Verlag, Berlin, 1973. xvi+388 pp. ISBN 3-540-06008-1.

[261] G. W. Stewart. *Introduction to Matrix Computations.* Academic Press, New York, 1973. xiii+441 pp. ISBN 0-12-670350-7.

[262] Gilbert Strang. *Introduction to Applied Mathematics.* Wellesley-Cambridge Press, Wellesley, MA, USA, 1986. ix+758 pp. ISBN 0-9614088-0-4.

[263] William Strunk, Jr. and E. B. White. *The Elements of Style.* Third edition, Macmillan, New York, 1979. xvii+92 pp. ISBN 0-02-418200-1.

[264] John Swales. *Writing Scientific English.* Thomas Nelson, London, 1971.

[265] John Swales. *Episodes in ESP: A Source and Reference Book on the Development of English for Science and Technology.* Prentice Hall International, Hemel Hempstead, Hampshire, UK, 1988. ISBN 0-13-283383-2.

[266] Michael Swan. *Practical English Usage.* Second edition, Oxford University Press, 1995. xxx+658 pp. ISBN 0-19-431197-X.

[267] Ellen Swanson. *Mathematics into Type: Copy Editing and Proofreading of Mathematics for Editorial Assistants and Authors.* Revised edition, American Mathematical Society, Providence, RI, USA, 1979. x+90 pp. ISBN 0-8218-0053-1.

[268] J. J. Sylvester. Explanation of the coincidence of a theorem given by Mr Sylvester in the December number of this journal, with one stated by Professor Donkin in the June number of the same. *Philosophical Magazine,* (Fourth Series) 1:44–46, 1851. Reprinted in [269, pp. 217–218].

[269] *The Collected Mathematical Papers of James Joseph Sylvester,* volume 1 (1837–1853). Cambridge University Press, 1904. xii+650 pp.

[270] Judith A. Tarutz. *Technical Editing: The Practical Guide for Editors and Writers.* Addison-Wesley, Reading, MA, USA, 1992. ISBN 0-201-56356-8.

[271] John Meurig Thomas. *Michael Faraday and the Royal Institution.* Adam Hilger, Bristol, UK, 1991. xii+234 pp. ISBN 0-7503-0145-7.

[272] Robert C. Thompson. Author vs. referee: A case history for middle level mathematicians. *Amer. Math. Monthly,* 90(10):661–668, 1983.

[273] Martin Tompa. Figures of merit. Research Report RC 14211 (#63576), IBM Thomas J. Watson Research Center, Yorktown Heights, New York, November 1988.

[274] Jerzy Trzeciak. *Writing Mathematical Papers in English: A Practical Guide.* Gdańsk Teachers' Press, Gdańsk, Poland, 1993. 48 pp. ISBN 83-85694-02-1.

[275] Edward R. Tufte. *The Visual Display of Quantitative Information.* Graphics Press, Cheshire, CT, USA, 1983. 197 pp.

[276] Edward R. Tufte. *Envisioning Information.* Graphics Press, Cheshire, CT, USA, 1990. 126 pp.

[277] Edward R. Tufte. *Visual Explanations: Images and Quantities, Evidence and Narrative.* Graphics Press, Cheshire, CT, USA, 1997. 158 pp. ISBN 0-9613921-2-6.

[278] Kate L. Turabian. *A Manual for Writers of Term Papers, Theses, and Dissertations.* Sixth edition, The University of Chicago Press, Chicago and London, 1996. ix+308 pp. ISBN 0-226-81627-3.

[279] Christopher Turk. *Effective Speaking: Communicating in Speech.* E & FN Spon (Chapman and Hall), London, 1985. ix+275 pp. ISBN 0-419-13030-6.

[280] Christopher Turk and John Kirkman. *Effective Writing: Improving Scientific, Technical and Business Communication.* Second edition, E & FN Spon (Chapman and Hall), London, 1989. 277 pp. ISBN 0-419-14660-1.

[281] Barry T. Turner. *Effective Technical Writing and Speaking*. Second edition, Business Books, London, 1978. xiii+220 pp. ISBN 0-220-66344-0.

[282] Adrian Underhill. *Use Your Dictionary: A Practice Book for Users of Oxford Advanced Learner's Dictionary of Current English and Oxford Student's Dictionary of Current English*. Oxford University Press, 1980. 56 pp. ISBN 0-19-431104-X.

[283] Mary-Claire van Leunen. *A Handbook for Scholars*. Revised edition, Oxford University Press, New York, 1992. xi+348 pp. ISBN 0-19-506954-4.

[284] Charles F. Van Loan. *Computational Frameworks for the Fast Fourier Transform*. Society for Industrial and Applied Mathematics, Philadelphia, PA, 1992. xiii+273 pp. ISBN 0-89871-285-8.

[285] Charles F. Van Loan. FFTs and the sparse factorization idea (abstract). *Linear Algebra and Appl.*, 162-164:717, 1992.

[286] Jan Venolia. *Write Right! A Desk Drawer Digest of Punctuation, Grammar and Style*. David St John Thomas Publisher, Nairn, Scotland, 1986. 126 pp. ISBN 0-946537-57-7.

[287] Keith Waterhouse. *On Newspaper Style*. Viking, London, 1989. 250 pp. ISBN 0-670-82626-X.

[288] Keith Waterhouse. *English our English (and How to Sing It)*. Viking, London, 1991. xxvii+147 pp. ISBN 0-670-83269-3.

[289] David S. Watkins. *Fundamentals of Matrix Computations*. Wiley, New York, 1991. xiii+449 pp. ISBN 0-471-54601-1.

[290] J. D. Watson and F. H. C. Crick. Molecular structure of nucleic acids: A structure for deoxyribose nucleic acid. *Nature*, 171(4356):737–738, 1953. April 25.

[291] Robert L. Weber, editor. *More Random Walks in Science*. Bristol and London, 1982. xv+208 pp. ISBN 0-85498-040-7.

[292] Robert L. Weber, editor. *Science with a Smile*. Institute of Physics Publishing, Bristol and Philadelphia, 1992. 452 pp. ISBN 0-7503-0211-9.

[293] *Webster's New World College Dictionary*. Third edition, Macmillan, New York, 1997. xxxvi+1588 pp. ISBN 0-02-861674-X.

[294] *Webster's Third New International Dictionary of the English Language*. Merriam-Webster, Springfield, MA, USA, 1986. 110+2662 pp. ISBN 0-87779-201-1, 0-87779-206-2.

[295] Herbert S. Wilf. TeX: A non-review. *Amer. Math. Monthly*, 93:309–315, 1986.

[296] J. H. Wilkinson. *The Algebraic Eigenvalue Problem*. Oxford University Press, 1965. xviii+662 pp. ISBN 0-19-853403-5 (hardback), 0-19-853418-3 (paperback).

[297] L. Pearce Williams, editor. *The Selected Correspondence of Michael Faraday: Volume 1, 1812–1848*. Cambridge University Press, 1971. ISBN 0-521-07908-X.

[298] Frederick T. Wood, Roger H. Flavell, and Linda M. Flavell. *Current English Usage*. Macmillan, London, 1989. v+329 pp. ISBN 0-333-27840-2.

[299] F. Peter Woodford. Sounder thinking through clearer writing. *Science*, 156 (3776):743–745, 1967.

[300] F. Peter Woodford, editor. *Scientific Writing for Graduate Students: A Manual on the Teaching of Scientific Writing*. Council of Biology Editors, Bethesda, MD, USA, 1986. x+187 pp. ISBN 0-914340-06-9.

[301] John D. Woolsey. Combating poster fatigue: How to use visual grammar and analysis to effect better visual communications. *Trends in Neurosciences*, 12(9):325–332, 1989.

[302] William Zinsser. *Writing with a Word Processor*. Harper and Row, New York, 1983. viii+117 pp. ISBN 0-06-091060-7.

[303] William Zinsser. *Writing to Learn*. Harper and Row, New York, 1988. x+256 pp. ISBN 0-06-091576-5.

[304] William Zinsser. *On Writing Well: An Informal Guide to Writing Nonfiction*. Fourth edition, HarperCollins, New York, 1990. xiii+288 pp. ISBN 0-06-272027-9.

Name Index

A suffix "t" after a page number denotes a table, "f" a figure, "n" a footnote, and "q" a quotation at the opening of a chapter.

Subject Index

At the laundress's at the Hole in the Wall in Cursitor's Alley
up three pair of stairs...
you may speak to the gentleman, if his flux be over,
who lies in the flock bed,
my index maker.

— DEAN SWIFT[27], A Further Account of the Most
Deplorable Conditions of Mr Edmund Curll, Bookseller,
Since His Being Poisoned on the 28th March (1716)

---

[27]Quoted in [156].

A suffix "t" after a page number denotes a table, "f" a figure, "n" a footnote, and "q" a quotation at the opening of a chapter. Definitions of technical terms are found in the glossary (Appendix E), which is not indexed here.